Physics Math Reset

Small Things and Vast Effects

Thomas J. Buckholtz

T. J. Buckholtz & Associates
Portola Valley, California
USA

Physics Math Reset
Small Things and Vast Effects

Edition 1

At the time this edition was published,

- The author could be reached via e-mail at Thomas.J.Buckholtz@gmail.com.
- The author had a website at http://thomasjbuckholtz.wordpress.com.

Library of Congress Control Number: 2014911029
CreateSpace Independent Publishing Platform, North Charleston, SC

Printed by CreateSpace Independent Publishing Platform, North Charleston, SC

Table of Contents

Preface

Welcome to *Physics Math Reset: Small Things and Vast Effects.*

In this book, I propose physics theory. Some of the theory falls beyond traditional physics research. Also, I discuss possibly new mathematics. I hope some of the concepts will prove useful.

You can use this book to gain non-traditional interpretations of nature. You can use concepts presented herein to think about fundamental physics and new applications of mathematics. Perhaps you will gain new vantage points for addressing traditional issues. Possibly you will see opportunities to enhance on-going or planned work. Perhaps you will find concepts for other research. Perhaps you will try to verify, refute, or extend work herein.

About the title

The following contributed to my choosing the title for this book.

- This book discusses attempted physics research. Hence, physics.
- This book discusses attempted math research. Hence, math.
- This book discusses an attempt to use new or underutilized math to model physical phenomena for which traditional math models fall short. Hence, reset.
- Hence, *Physics Math Reset.*
- This book features an attempt to develop models correlating with elementary particles and their properties. Hence, small things.
- This book features an attempt to develop models correlating with the rate of expansion of the universe and the shape of the universe. Hence, vast effects.
- Hence, *Physics Math Reset: Small Things and Vast Effects.*

About my hopes

I hope people will use this work. I hope people will benefit from this work. I hope people will tell me of extensions to this work, shortcomings in the work, and developments to which the work contributes.

- Thomas J. Buckholtz

Portola Valley, California USA
June 2014

Dedication

To Helen Buckholtz
And, in memory of Joel and Sylvia J. Buckholtz

About the author

Dr. Thomas J. Buckholtz is the author or a coauthor for articles, books, chapters, or reports regarding physics, applied physics, mathematics, computer science, applied computing, computer-based games, software licensing, innovation, systems-thinking tools, the information age, information proficiency, service science, governmental service to the public, and the role of chief information officers.

He played pivotal roles in the following endeavors.

- Create lines of business for a $1 billion (annual revenue) business unit.
- Save $100 million per year for a $6 billion company.
- Pioneer three information technologies.
- Establish three information-technology marketplace business practices.
- Develop useful business, engineering, and scientific software.
- Double a two-person firm's revenue, for each of two consecutive years.
- Preserve 7 kilometers of Pacific Ocean coastline.
- Create an international service program.
- Improve governmental service (from all levels of government) for the American public.
- Create a grassroots line-of-business for a United States political party's National Committee.

Tom served in the following capacities.

- Executive leading a $1 billion business unit
- Corporate officer and advisor for startups
- Chief information officer (CIO) for a $10 billion enterprise
- Co-CIO for the United States federal government's Executive Branch
- Program leader advocating innovation, enhancing teamwork, and providing information technology throughout a $6 billion company
- Commissioner, United States General Services Administration
- Mathematician; Scientist; Engineer
- Professorial Lecturer; University Extension Instructor
- Speaker; Workshop provider; Author
- Business advisor; Innovation consultant

Dr. Buckholtz's clients and employers have included large and small enterprises in aerospace, agricultural research, biotech, business services, computing, defense, education, energy utilities, government, healthcare, high technology, innovation, insurance, Internet, law enforcement, politics, research and development, telecommunications, and venture capital.

Tom served on elected or appointed boards or in other volunteer capacities for a residential cooperative, a swim club, and organizations in academia, innovation, and public policy. For a successful United States presidential campaign, he served as a donor, fundraiser, policy-research committee member, speaker, alternate delegate at the candidate's party's National Convention, speakers bureau leader, county cochairman, and county representative at regional and statewide meetings. He served as co-producer and co-host for 250 interview-format television programs discussing business, charitable, community, educational, governmental, and political endeavors.

His education includes the following.

- Earn a B.S. in mathematics from the California Institute of Technology.
- Earn a Ph.D. in physics from the University of California, Berkeley.
- Complete business administration programs at Stanford University and the University of Michigan.

Notice

Small Things and Vast Effects

Thomas J. Buckholtz

Thomas.J.Buckholtz@gmail.com

In particle theory and cosmology theory, some problems have remained unsolved for decades. For example, predict remaining undiscovered basic particles. Describe quantum gravity simply. Explain changes in the rate of expansion of the universe.

Why have people not solved these problems? People add quantum models on top of classical-physics math.

What do we do? We try a new approach. We start from quantum experimental results. We develop IOM (quantum isotropic harmonic oscillator methods, math, and models).

We use the models to catalog basic particles - some known and some we predict - with $0 \leq S \leq 5$. (S denotes spin/ħ.) We point to possible charges and approximate masses for basic bosons we predict. We state a finite set of possible mass-math-eigenvalues or masses for neutrinos. We predict a smaller than observed uncertainty range for the tauon mass.

Solved problems may include the following. Describe a symmetry that encompasses Standard Model symmetries. Describe quantum gravity. Describe forces that govern the rate of expansion of the universe. Describe forces that led to the universe being flat. Describe candidates for dark matter. Describe candidates for the stuff that contributes the density of the universe that people sometimes attribute to dark energy. Describe reactions that led to matter/antimatter imbalance. Describe reactions that lead to neutrino oscillations. Point to contributions to C, P, CP, and T violations, beyond contributions with which the Standard Model correlates.

The models correlate with 3 generations of fermions. People might say that the models unify all basic forces - known and predicted. People can use IOM to obviate concerns about possibly infinite zero-point vacuum energy.

Possibly, people can further use IOM. IOM include underutilized solutions to well-known equations.

Keywords

(Traditional terms:)

- Asymptotic freedom
- Axion
- Baryonic matter
- Black hole
- Boson
- Charge
- Clumping
- Color charge
- Cosmic microwave background (CMB)
- CP violation
- CPT symmetry
- Curvature of the universe
- Dark energy
- Dark matter
- Density of the universe
- Elementary particles
- Entanglement
- Fermion
- Flat universe
- Fundamental forces
- Generations
- Gluon
- Graviton
- Group larger than SU(3)×SU(2)×U(1)
- Higgs boson
- Lasing
- Leptoquark
- Magnetic moment

- Mass
- Masses of elementary particles
- Matter/antimatter imbalance
- Neutrino masses
- Photon
- Quantum gravity
- Quantum harmonic oscillator
- Quasar
- P violation
- Rate of expansion of the universe
- Space-like behavior
- Spin
- Standard Model
- Standard Model symmetry
- Strong interaction
- Time-like behavior
- SU(3)×SU(2)×U(1)
- Unified electromagnetism and gravity
- Vector potential
- Weak interaction

(Non-traditional terms:)

- Channel
- Coherence
- Ensemble
- Quantum interaction space

Part 1 Introduction

Section 1 Context

Abs.1.1 Traditional mathematical models do not adequately correlate with physics observations.
Abs.1.2 We develop models based on quantum observations and harmonic-oscillator math.

We note aspects of physics for which mathematical models fall short

People develop mathematical models that correlate with physics observations. People use such models to talk about past observations and to predict future observations.

The next items list needs for which people say mathematical models fall short. The term basic particle denotes a particle (such as an electron) that we think is not composed of other particles. We use the term compound to denote particles (such protons) that include more than 1 basic particle.

Unmet needs	(1.1)
Provide models people can use to ...	(1.2)
• Advance elementary-particle physics	(1.3)
• Provide a basis for the Standard Model symmetry SU(3)×SU(2)×U(1)	(1.4)
• Determine physics-relevant groups that contain SU(3)×SU(2)×U(1)	(1.5)
• List possible basic particles that have not been observed	(1.6)
• Explain the number, 3, of generations of fermions	(1.7)
• Describe quantum gravity	(1.8)
• Unify quantum gravity and quantum electromagnetism	(1.9)
• Explain the sizes of some symmetry violations (P, CP, ...)	(1.10)
• Explain neutrino oscillations	(1.11)
• Predict neutrino masses	(1.12)
• Interrelate masses of basic particles other than charged leptons	(1.13)
• Describe dark matter	(1.14)
• Describe dark energy	(1.15)
• Advance cosmology	(1.16)
• Explain changes in the rate of expansion of the universe	(1.17)
• Explain the flatness of the universe ($\Omega_0 \approx 1$)	(1.18)
• Advance other understanding	(1.19)
• Explain baryon asymmetry (matter/antimatter imbalance)	(1.20)
• Address the zero-point energy of the vacuum	(1.21)

We contrast traditional mathematical physics and our approach

The next item notes features of traditional approaches. For example, people say that people base the Standard Model for elementary particles on math that includes Lagrangian math.

Traditional approaches: (1.22)

- Start from classical-physics experimental results
- Develop math to model classical-physics experimental results
  - For example, use Lagrangian math
- Observe quantized phenomena
- Add quantum math on top of classical-physics math
- Look at the extents to which ...
  - Models correlate with known experimental and observational physics
  - Models predict new physics

The next item notes features of our approach.

Our approach: (1.23)

- Start from quantum-physics experimental results
- Develop math to model quantized phenomena
  - We use the term IOM to denote quantum isotropic harmonic oscillator methods, math, and models
  - We use quantum-physics and classical-physics experimental results
- Look at the extents to which ...
  - IOM correlate with known experimental and observational physics
  - IOM predict new physics

We discuss characteristics of our approach

The next items list some characteristics of our approach.

- Feature 1 math basis (1.24)
- Include and identify guesses (1.25)
- State finite sets of choices, when we do not decide between choices (1.26)
- Anticipate that further work can ... (1.27)
  - Describe the math (and overall attempted research) more completely and more easily
  - Convert guesses to better-supported assumptions
  - Get more results from the math
  - Reduce the number of choices in each of some sets of choices
- Anticipate that further work can ... (1.28)
  - Provide a substrate from which people rebuild the Standard Model for elementary particles
  - Provide a substrate from which people gain insight regarding the uncertainty principle
  - Provide a substrate from which people rebuild classical mechanics and classical electrodynamics
  - Provide a substrate from which people explore aspects governing the early evolution of the universe

The next items provide examples of item (1.26). Item (1.29) dovetails with a traditionally unresolved matter as to whether neutrinos are their own antiparticles. [Section 11] Item (1.30) illustrates a possible improvement over traditional results, which do not specify how to compute neutrino masses. [Section 18]

- We identify 2 possibilities for neutrinos (1.29)
  - Each of the 3 neutrinos is its own antiparticle
  - Each neutrino has a distinct antiparticle
- We identify and approximately calculate 4 masses, of which no more than 3 would be candidate masses for neutrinos or candidate mass-related eigenvalues for neutrinos (1.30)

Section 2 Core

Abs.2.1 IOM (quantum isotropic harmonic oscillator methods) may correlate with and predict physics observations with which traditional models do not correlate.

We discuss some math relevant to our approach

The next items list some math we correlate with quantum-physics experimental results. Each item features a set of discreet numbers (and not a continuous range of numbers).

- For each known basic particle, the expression $ï|q_e|/3$, with ï being an integer, describes the charge of the particle (2.1)
  - q_e denotes the charge of an electron
- For each basic particle, the expression $S\hbar$, with 2S being a non-negative integer, describes the spin of the particle (2.2)
  - $\hbar$ denotes Planck's constant (reduced)
- Harmonic-oscillator-math raising operators and lowering operators correlate with quantum descriptions of modes of the vector potential (2.3)
  - The spectrum of excitation states is discreet
- The number $S(S+D_{*P}-2)$, with $D_{*P}=3$, pertains for the radial-component of some harmonic-oscillator math correlating with particles having spin/$\hbar$ of S (2.4)
  - $S(S+D_{*P}-2) = S(S+1)$, for $D_{*P}=3$
  - For each known basic particle, people consider such an $S(S+1)$ to be relevant to the physics of the particle
- For each basic particle, a non-negative number m, describes the mass of the particle (2.5)
  - The spectrum of such masses is discreet
- The single speed c and the single expression $E^2-c^2P^2=m^2c^4$ pertain for all free-ranging basic particles (2.6)
  - In the expression, E denotes energy, c denotes the speed of light (in a vacuum), P denotes momentum, and m denotes the mass of the specific particle
  - Free-ranging particles include electrons and photons
  - Free-ranging particles do not include quarks
- For each known basic fermion, the number of generations is 3 (2.7)

People might say that item (2.3) pertains exactly, at least to the extent of known photonics. (People might say that other uses of harmonic-oscillator math in traditional physics feature approximations to physics phenomena. For example, attempts to quantize aspects of classical physics exactly via harmonic-oscillator math would require modelling an infinitely large potential energy.)

Regarding item (2.4), people might say that $D_{*P}=3$ correlates with the usefulness of people discussing a concept of a (perhaps curved) space time with 3 spatial dimensions. (For completeness, we note that, for D an integer and D≠3, applications of harmonic-oscillator math exist for which an analog to S(S+D−2) pertains. We make use of some D≠3 applications, as well as of applications for which D=3.)

We anticipate a prediction based on IOM and we note a contrast with traditional results

IOM correlate with possible basic particles for which S can be as much as 5. A subset of IOM correlates with all known basic particles.

In contrast, the Standard Model features basic particles for which S does not exceed 1.

We note possible usefulness of IOM

IOM seem to provide new insight regarding items following item (1.1). Items following item (24.1) estimate extents to which work in this paper provides new insight regarding items following item (1.1). Perhaps people would say that IOM solve - or point significantly toward solutions for - each item following item (1.1).

IOM seem to provide other new insight.

We are not aware of contradictions between observational physics and results of IOM.

Perhaps, people would say that this paper does not completely describe or use IOM. We would agree. [Items including and following items (1.24), and items including and following item (1.29)]

Perhaps, when fully developed, IOM will provide adequate insight regarding what people now consider to be fundamental physics.

Section 3 Comments

Abs.3.1 IOM include a concept - QI space - that people might say merges some concepts regarding quantum mechanics, energy-momentum space, and space time.

We discuss consistency and accuracy of work this paper presents

The next items provide observations we think people might make about work in this paper.

- The work features related models (3.1)
  - The models reinforce each other
  - The models do not contradict each other
  - Perhaps, people can produce a more unified treatment
- Some predictions may be incomplete or wrong (3.2)
  - For example, each of some models leads to finite choices for some physical numbers, but not to a single prediction for each number
- Removal of incorrect work or predictions need not lead to overturning significantly other work or predictions (3.3)

The next items provide an example of item (3.1). Each of the next 2 items points toward the notion that S for observable zero-mass basic bosons does not exceed 4. For item (3.4), the integers 4, 3, 2, and 1 correlate respectively with S = 1, 2, 3, and 4. The integer 0 would correlate with an inability to detect effects of a perhaps otherwise-possible zero-mass basic boson with S=5. [Section 6 and Section 8] For item (3.5), if the SU(9) possibility pertains, the SU(9) considerations might pertain to the same perhaps otherwise-possible zero-mass basic boson with S=5. [Section 13]

- A factor of 4/3 appears in an algebraic equation that may link ... (3.4)
  - The ratio of the mass of a tauon to the mass of an electron
  - A ratio of strengths of electromagnetism and gravity
- A property of a series of apparently relevant integers, each correlating with an application of a group SU(ï), changes as ï transits from 7 to 9 (3.5)
  - The integers are 8, 24, 48, and 80
  - The integers correspond to numbers of generators associated with SU(ï) for ï = 3, 5, 7, and 9, respectively
  - The numbers would correlate respectively with S = 2, 3, 4, and 5
  - We think that this application of the series (8, 24, 48 ...) ends at the last integer for which division by the preceding integer in the series produces an integer
  - We interpret IOM as correlating with possible existence of an S=4 zero-mass boson (48/24 is an integer)
  - We interpret IOM as correlating with non-existence of an S=5 zero-mass basic boson (80/48 is not an integer)

People might say that this paper does not directly show a root adequately common to items (3.4) and (3.5). Perhaps, people will someday better show links between items (3.4) and (3.5).

We discuss QI space and concepts related to space time and to energy-momentum space

In traditional physics, people use concepts of space time and concepts of energy-momentum space.

People might say that IOM correlate with a quantum hybrid of space time and energy-momentum space. We use the term QI space. Here, QI abbreviates the phrase quantum interaction. We feature the term quantum because IOM feature quantum physics. The next items illustrate why we use the term hybrid.

People might say that ... (3.6)

- Aspects of QI space seem to correlate with a quantum version of energy-momentum space (3.7)
- Aspects of QI space seem to correlate with a (space time) direction of motion for photons (3.8)

In traditional physics, people apply terms such as time-like to the separation of 2 points (in space time) for which (relative to a geodesic connecting the points) the difference in (the speed of light multiplied by) time exceeds the spatial difference. In traditional physics, people apply terms such as space-like to the separation of 2 points (in space time) for which (relative to a geodesic connecting the points) the spatial difference exceeds the difference in (the speed of light multiplied by) time.

For convenience, we use the terms the next items list. We apply QE-like and QP-like to discussions of QI space, space time, and energy-momentum space.

Our terminology	More-traditional terminology	
		(3.9)
QE-like	• time-like (regarding space time) • energy-like (regarding energy-momentum space)	(3.10)
QP-like	• space-like (regarding space time) • momentum-like (regarding energy-momentum space)	(3.11)

We discuss roles for guesses

This paper includes guesses.

The following statement provides a description of how science works made by Richard Feynman. [Ref.3.1]

> In general we look for a new law by the following process. First we guess it. Then we compute the consequences of the guess to see what would be implied if this law that we guessed is right. Then we compare the result of the computation to nature, with experiment or experience; compare it directly with observation, to see if it works. If it disagrees with experiment it is wrong.

An approach to the research in this paper features activities that include the following. Guess at a concept for an area to explore. Think of possible patterns or numeric relationships. Develop tentative models or theory. Review models or theory with respect to observations or experiments, traditional vocabulary and statements, and aspects of this work. Write tentative results into a draft paper covering aspects of the research. Review the research for consistency. Review the draft for readability and consistency. Go back to fix problems. Go forward regarding other areas.

We discuss summaries and compendia this paper contains

Items following item (24.1) estimate extents to which work in this paper provides new insight regarding items following (1.1).

The next items note places where this paper performs various functions. The place column indicates the section that performs the activity the function column notes. The list column shows letters that partly label some statements in this paper. This paper includes statements with labels of the form ύ.ϊ.ό for which ύ is 3 letters, ϊ is a section number, and ό is a number. For example, the 3 letters Abs denote abstract. Statements labeled in the form Abs.ϊ.ό summarize contents of sections. This paper lists references regarding numeric data.

Function	Place	List	
			(3.12)
Summarize results	Section 25	Abs	(3.13)
List guesses	Section 25	Gss	(3.14)
List possible opportunities for observational or experimental research	Section 25	SOR	(3.15)
List possible opportunities for theoretical research	Section 25	STR	(3.16)
List references	Section 26	Ref	(3.17)

We discuss structural elements this paper uses

Starting with Part 2, each section before Part 6 contains up to 4 elements - section abstract, context, core, and comments. Starting with Part 2, the introduction to each part before Part 6 contains 2 elements - context and core.

We list references

Ref.3.1 John R. Gribbin and Mary Gribbin, *Richard Feynman, A Life In Science*, Dutton, 1997, page 178.

Part 2 Some math, physics, and numbers

Context

We discuss traditional mathematics and traditional physics

People describe, via traditional mathematics, math and solutions for isotropic harmonic oscillators.

People describe, via traditional applications of Lagrangian math, models people correlate with elementary particles. People use the term Standard Model to describe some of this work.

We anticipate math and models correlating with aspects of basic particles

We anticipate discussing underutilized math and solutions for isotropic harmonic oscillators. We anticipate offering math related to isotropic harmonic oscillators as an approach to cataloging basic particles and to describing some properties of those particles.

Core

We preview sections in this part

Section 4 discusses math for quantum isotropic harmonic oscillators. We find possibly underutilized solutions. We provide possibly new notation.

Section 5 builds bridges between models based on quantum isotropic harmonic oscillators and people's observations regarding nature.

Section 6 section reviews some numbers people measure and defines some numbers we use.

Section 4 IOM (quantum isotropic harmonic oscillator methods, math, and models)

Abs.4.1 We introduce IOM.

Context

We note traditional applications of harmonic-oscillator math

People use harmonic-oscillator math to model the motion of a mass attached to a spring. People state quantum versions of such math. People use quantum harmonic-oscillator math to model interactions between objects. Such applications do not pertain to modeling basic properties (such as charge or mass) of basic particles (such as electrons or quarks). People use quantum harmonic-oscillator math to discuss the vector potential and to discuss lasing. As far as we know, people tend not to use negative quantum numbers for quantum harmonic-oscillator math.

We anticipate some uses of harmonic-oscillator math

We correlate quantum-harmonic oscillator math with internal (or, invariant) properties (such as charge or mass) of basic particles. For non-zero-mass basic particles, such applications model mass (for bosons) and abilities to interact via bosons (for fermions). For zero-mass particles, such applications model concepts including and paralleling the vector potential.

Sometimes, we do not solve for an energy-like number. We assume that such a quantity exists or is not relevant. We look for symmetries.

Core

We note some types of traditional quantum-mechanics representations

People provide math to describe amplitudes for quantum states. The next items list 2 approaches.

A QM-type-C approach features terms expressed as functions (wave functions) of spatial coordinates (4.1)
- Some such wave functions satisfy partial differential equations

A QM-type-D approach uses terms not expressed by using spatial coordinates (4.2)
- Some such terms satisfy conditions related to raising operators and lowering operators

People can base use of each approach on either of 2 types of coordinates.

CO-type-L features linear coordinates (4.3)
- For 3-dimensional quantum mechanics, people might denote coordinates by x, y, and z

CO-type-S features radial and angular coordinates (4.4)
- For 3-dimensional quantum mechanics, people might denote the radial coordinate by r, with $r=(x^2+y^2+z^2)^{1/2}$

The next items show typical coordinates people use. These coordinates apply for 3 spatial dimensions.

(4.5)

Coordinate system	Coordinates	Symmetry point	Distance from symmetry point
CO-type-L	x, y, z	x=y=z=0	$r=(x^2+y^2+z^2)^{1/2}$
CO-type-S	r, θ, φ	r=0	r

$-\infty<ύ<\infty$, for ύ = x, y, or z (4.6)

$0\le r<\infty$, $0\le\theta\le\pi$, and $0\le\varphi\le 2\pi$ (4.7)

We note approaches we use

The next items show paired approaches we use.

QM-type-DL denotes an approach that combines QM-type-D and CO-type-L (4.8)

QM-type-CS denotes an approach that combines QM-type-C and CO-type-S (4.9)

QM-type-CL denotes an approach that combines QM-type-C and CO-type-L (4.10)

We denote 2 numbers of dimensions. Some people may choose to associate the subscript E with the terms energy-like or time-like. We use the term QE-like. Some people may choose to associate the subscript P with the terms momentum-like or space-like. We use the term QP-like.

D_E denotes an integer, with $D_E>0$	(4.11)
D_P denotes an integer, with $D_P>0$	(4.12)

The next items denote 2 sets. Each set consists of indices associated with a sequence of consecutive integers.

$$\text{SIDE} = \{ E\ddot{\imath} \mid \ddot{\imath} = D_E, D_E-1, D_E-2, \ldots, 3, 2, \text{or } 1 \} \quad (4.13)$$
$$\text{SIDP} = \{ P\ddot{\imath} \mid \ddot{\imath} = 1, 2, 3, \ldots, D_P-2, D_P-1, \text{or } D_P \} \quad (4.14)$$

The next item denotes the union of the 2 sets.

$$\text{SID} = \text{SIDE} \cup \text{SIDP} \quad (4.15)$$

We use IOM for which the next items pertain. Here, for item (4.17), we use QM-type-DL. Here, $\in$ denotes belongs to (or, is a member of). Here, n_χ denotes the quantum number for harmonic oscillator χ. Each n_χ + 1/2 term correlates with a result from traditional harmonic-oscillator math. Regarding item (4.18), the term isotropic correlates with each $\pm_\chi$ having the same magnitude as each other $\pm_\chi$. The difference in signs (+1 vis-à-vis −1) has significance. The reverse choice of signs can also work.

$$D_E \text{ is odd and } D_P \text{ is odd} \quad (4.16)$$
$$0 = Œ = \Sigma_{\chi\in\text{SID}} \pm_\chi(n_\chi + 1/2) \quad (4.17)$$
$$\pm_\chi = +1 \text{ for } \chi\in\text{SIDE} \quad (4.18)$$
$$\pm_\chi = -1 \text{ for } \chi\in\text{SIDP}$$

The next item repeats a feature of item (4.17).

$$Œ = 0 \quad (4.19)$$

We show a traditional solution

The next items pertain to IOM for a traditional ground state of an isotropic harmonic oscillator with 3 spatial dimensions. Item (4.22) contributes +3/2 to item (4.17). Item (4.23) contributes −3/2 to item (4.17).

$$D_E = 1 \quad (4.20)$$
$$D_P = 3 \quad (4.21)$$
$$n_{E1} = 1 \quad (4.22)$$
$$n_{P1} = n_{P2} = n_{P3} = 0 \quad (4.23)$$

We explore aspects of a QM-type-CS approach and find non-traditional solutions

The next items show math for a QM-type-CS approach. Ψ can be a function also of coordinates (angular coordinates) other than r. People call item (4.25) the Laplacian operator for D dimensions. People call item (4.32) the potential.

$$\xi\, \Psi(r) = (\xi_0/2)\, (\, -\eta^2\, \nabla^2 + \eta^{-2} r^2\,)\, \Psi(r) \quad (4.24)$$
$$\nabla^2 = r^{-(D-1)}(\partial/\partial r)(r^{D-1})(\partial/\partial r) - \Omega r^{-2} \quad (4.25)$$

ξ and $\xi_0/2$ denote numbers	(4.26)

$\Psi(r)$ denotes a wave function (4.27)
r denotes a variable, with dimensions of length (4.28)
η denotes a length (4.29)
Ω denotes a number (4.30)
D denotes a non-negative integer (4.31)

$$V = (\xi_0/2)\, \eta^{-2}\, r^2 \quad (4.32)$$

The next item pertains.

For D=1, some solutions feature the following (4.33)

- Ω=0
- The range $-\infty < r < \infty$ pertains
- ψ has the form of a Hermite polynomial (in variable r) multiplied by $\exp(-r^{-2}/(2\eta^2))$

Work below tends not to use solutions people associate with (4.33). Below, the range $0 \leq r < \infty$ pertains for traditional solutions.

The next item describes solutions other than solutions people traditionally associate with (4.33).

$$\psi(r) \propto r^{\nu} \exp(\, -r^{-2} \,/\, (2\eta^2)\,) \quad (4.34)$$

The next items pertain for all solutions for which item (4.33) does not pertain. The parameter η does not appear in these items.

$$\xi = (D+2\nu)\, (\xi_0/2) \quad (4.35)$$
$$\Omega = \nu(\nu+D-2) \quad (4.36)$$

The next items pertain to traditional solutions.

ν is non-negative (4.37)
ν is an integer (4.38)
Ω is non-negative (4.39)

Each of the next items points to non-traditional solutions.

ν can be negative (4.40)
ν can be other than an integer (4.41)

For D>2, item (4.40) is necessary (but not sufficient) for the next item to pertain.

Ω can be negative (4.42)

We focus on solutions that normalize

We limit our attention to solutions that can be normalized.

The next item shows behavior of the r-related normalization integrand near r=0.

$$\Psi^*\Psi\, r^{D-1} \sim r^{D-1+2\nu} \exp(-2r^2 2^{-1}\eta^{-2}) \sim r^{D-1+2\nu}, \text{ for } r{\sim}0 \quad (4.43)$$

The next items pertain to some solutions that normalize.

$$-1 < D-1+2\nu \quad (4.44)$$
$$-D/2 < \nu \quad (4.45)$$
$$\psi \text{ normalizes if (but not only if) ... } -(D/2) < \nu \quad (4.46)$$

The next item provides a definition of the Dirac delta function. [Ref.4.1]

$$\delta(r) = \lim_{\varepsilon\to 0+} (1/(2(\pi\varepsilon)^{1/2})) \exp(-r^2/(4\varepsilon)) \quad (4.47)$$

We make the following association.

$$4\varepsilon = \eta^2 \quad (4.48)$$

We assume that use of items (4.47) and (4.48) correlates with extending the range of integration. The next item shows an extended range of integration. Perhaps, people should consider that $r_E=\infty$.

$$-r_E \leq r < \infty \quad (4.49)$$
$$r_E > 0$$

For r<0, we note the possibility that the angular dependence of Ψ changes. For example, $\cos(\theta)$ for r>0 might become $\cosh(\theta)$ for r<0. We anticipate the possibility of products of exponentials and trigonometric functions.

The next item supplements item (4.46).

$$\psi \text{ normalizes if (but not only if) ... } -(D/2) = \nu \quad (4.50)$$

We coin these terms.

Inside denotes $-(D/2) < \nu$ (4.51)
Edge denotes $-(D/2) = \nu$ (4.52)

For each of inside or edge, η can have any real value other than 0. Two sets of mathematical solutions exist. One set corresponds to $\eta>0$. The other set corresponds to $\eta<0$.

For an edge case with -2ν an even integer, for each solution set, potentially 2 solutions exist. (For example, for $\Omega=0$, one potential solution has $\Psi(-r)=\Psi(r)$. Another potential solution has $\Psi(-r)=-\Psi(r)$.)

We call a linear combination (of potential solutions) that normalizes a type-1 solution. We call a linear combination that does not normalize a type-2 solution.

We base the following item on considerations related to item (4.49).

Gss.4.1 For an edge case with -2ν an even positive integer, 1 type-1 solution exists. (4.53)

For an edge case with -2ν an odd positive integer, 2 square roots of r^ν exist. Potentially, for each solution set, 4 solutions exist.

Gss.4.2 For an edge case with -2ν an odd positive integer, 3 orthogonal type-1 solutions exist. (4.54)

People apply the next items to traditional $D_P=3$, $r>0$ math.

$D_P = 3$	(4.55)
$D = 3$	(4.56)
$S = \nu$, for some non-negative integer ν	(4.57)
$\Omega = \nu(\nu+D-2) = S(S+D_P-2) = S(S+1)$	(4.58)
$2S+1$ angular solutions pertain	(4.59)

The next items extend traditional $D_P=3$ math. $D \neq D_P$ is allowed, as is $D=D_P$.

$D_P = 3$	(4.60)
$-D_P \leq 2\nu$, with 2ν being an integer	(4.61)
$\Omega = \nu(\nu+D-2)$, for some non-negative integer D	(4.62)
$\lvert\Omega\rvert = S(S+D_P-2) = S(S+1)$, for some S with 2S being a non-negative integer	(4.63)
$2S+1$ angular solutions pertain	(4.64)

We tabulate solution sets and solutions

The next items summarize results. Regarding the numbers of solutions sets, the leftmost factor of 2 comes from the existence of 2 cases, namely $\eta>0$ and $\eta<0$.

Type	-2ν	Number of solution sets	Orthogonal type-1 solutions per set	(4.65)
inside	even and >0	2(2S+1)	1	(4.66)
inside	odd and >0	2(2S+1)	1	(4.67)
edge	even and >0	2(2S+1)	1	(4.68)
edge	odd and >0	2(2S+1)	3	(4.69)

We provide notation for solution sets and solutions

The next items provide notation for solution sets.

The symbol `s± denotes a solution set	(4.70)
2s is an integer, with $-S \leq s \leq S$	(4.71)
± is + for $\eta>0$ and is − for $\eta<0$	(4.72)

The next items provide notation for type-1 solutions.

The symbol `s±ï denotes a type-1 solution	(4.73)
`s± denotes the solution set	(4.74)
ï is an integer, with $1 \leq$ ï $\leq$ the number of orthogonal type-1 solutions in the solution set	(4.75)

We show a non-traditional solution

We apply a QM-type-CS approach to the example that items including and following item (4.20) show. The next items show a non-traditional solution. This is an inside solution. Here, $S \neq \nu$.

$D_P = 3$	(4.76)

$$\nu = -1 \quad (4.77)$$
$$D = 3 \quad (4.78)$$
$$\Omega = \nu(\nu+D-2) = -1(0) = 0 \quad (4.79)$$
$$S = 0 \quad (4.80)$$
$$\Omega = S(S+1) \quad (4.81)$$

We consider this non-traditional solution to correlate with the ground state. The next items pertain.

$$\xi = (1/2)\ \xi_0 \quad (4.82)$$
$$\psi(r) \propto r^{-1} \exp(\ -r^{-2} / (2\eta^2)\ ) \quad (4.83)$$

Along with the non-traditional solution, the next items pertain. These solutions correlate with solutions items including and following item (4.20) suggest. Traditionally, people state that the S=0 solution below corresponds to the ground state. For the S=0 solution below, $\xi = (3/2)\ \xi_0$.

$$\xi = (D+2\nu)\ (\xi_0/2) = (3/2 + S)\ \xi_0 \quad (4.84)$$
S is a non-negative integer (4.85)
$$\Omega = S(S+1) \quad (4.86)$$
$$\psi(r) \propto r^S \exp(\ -r^{-2} / (2\eta^2)\ ) \quad (4.87)$$

We show another non-traditional solution

We provide an example of a non-traditional solution correlating with QM-type-DL representations.

Items including and following item (4.20) show a traditional ground state. Item (4.22) contributes +3/2 to item (4.17). The next items show this traditional ground state.

n_{E1}	n_{P1}	n_{P2}	n_{P3}	(4.88)
1	0	0	0	(4.89)

We consider non-traditional solutions. The next items show an n'-times excited state for the P2 oscillator. Here, n' denotes a non-negative integer.

n_{E1}	n_{P1}	n_{P2}	n_{P3}	(4.90)
n'	−1	n'	0	(4.91)

Here, n'=0 denotes the ground state. For n'=0, the n_{E1}-based contribution to item (4.17) is +1/2. The next items show this non-traditional ground state.

n_{E1}	n_{P1}	n_{P2}	n_{P3}	(4.92)
0	−1	0	0	(4.93)

We define open and closed, for pairs of harmonic oscillators

We use QM-type-DL. We consider 2 oscillators. One oscillator has index ό. The other oscillator has index ύ. Here ό ≠ ύ. Here, the state of the oscillator pair can be a sum of components. Each component consists of a product of a non-zero complex number and a basis amplitude we symbolize by $|\ n_{ό}\ , n_{ύ} \rangle$.

We assume exactly 1 of the following (4.94)

- ó∈SIDE and ύ∈SIDE
- ó∈SIDP and ύ∈SIDP

We say that the ó-and-ύ pair of oscillators is open if there is exactly 1 component and (for that component) each of $n_{ó}$ and $n_{ύ}$ is a single integer (4.95)

- Examples include
  - $n_{ó} \geq 0$, $n_{ύ} \geq 0$
  - $n_{ó}=-2$, $n_{ύ}=-1$
  - $n_{ó}=-1$, $n_{ύ}=-1$

We say that the ó-and-ύ pair of oscillators is closed if the state of the oscillator pair is a linear combination of at least 2 distinct components, each having $n_{ó} + n_{ύ} = -1$ (4.96)

- An example is
  - $(1/2)^{1/2}$ (| $n_{ó}=0$, $n_{ύ}=-1$ > + | $n_{ó}=-1$, $n_{ύ}=0$ >)

We use a single * to denote the state of an oscillator that participates in a closed pair.
The concept of closed allows, for example, considering solutions for differing D_P to be equivalent.
For example, we can consider the $D_E=1$, $D_P=5$ solution that item (4.98) shows to be equivalent to the $D_E=1$, $D_P=3$ solution that item (4.100) shows

n_{E1}	n_{P1}	n_{P2}	n_{P3}	n_{P4}	n_{P5}	(4.97)
1	0	0	0	*	*	(4.98)

n_{E1}	n_{P1}	n_{P2}	n_{P3}	(4.99)
1	0	0	0	(4.100)

For each of these 2 solutions, the P2-and-P3 pair is open.

We narrow the types of oscillator pairs we consider

For the remainder of this paper, we restrict the definition of oscillator pair (for other than the E1-and-P1 pair). The next item shows this restriction.

For an ó-and-ύ oscillator pair associated with SIDE, (4.101)

- ύ has the form Eï, in which ï is an even integer
- ó has the form Eϋ, in which ϋ=ï+1

For an ó-and-ύ oscillator pair associated with SIDP,

- ó has the form Pï, in which ï is an even integer
- ύ has the form Pϋ, in which ϋ=ï+1

Sometimes we consider the oscillator pair E1-and-P1.

Comments

We suggest research

STR.4.1 Complete mathematics related to type-1 solutions sufficiently to describe wave functions for edge cases for $D_P = 3$ and $\nu = -3/2$. (Possibly, extend the work to pertain to edge cases for other D_P and ν.)

STR.4.2 To what extent might people derive benefit from IOM for any 1 or more than 1 of the following? D can be an integer < 1. D can be other than an integer. 2ν can be other than an integer. $\pm_\chi$ can be other than +1 or −1.

STR.4.3 To what extent would it be useful, for D=3, for people to consider a quantum number s' such that $S=(s'-1)/2$? (Here, $S(S+1)=(1/4)\cdot((s')^2-1)$. Here, perhaps, s' is any non-zero integer.)

STR.4.4 Explore IOM for cases in which D_E is even and D_P is even.

STR.4.5 How might people improve or extend the technique for cataloging quantum approaches?

We list references

Ref.4.1 Wolfram Alpha, computational knowledge engine, Wolfram Alpha LLC, http://mathworld.wolfram.com/DeltaFunction.html.

Section 5 IOM and physics

Abs.5.1 We focus on IOM for which $\Omega = \pm S(S+1)$, with S=spin/ħ and with 2S being an integer.

Abs.5.2 IOM correlate with boson and fermion particles and fields.

Context

We note properties people associate with traditional physics

As far as we know, people interpret all experiments and observations as being consistent with the next items.

Each basic particle has a spin/ħ people can denote by S	(5.1)
For each basic particle, 2S is a non-negative integer	(5.2)
For each known basic particle, S = 0, 1/2, or 1	(5.3)
People expect that, if gravitons exist, S=2 for gravitons	(5.4)

We anticipate use of IOM

The next items characterize the set of S that we anticipate to be relevant.

For each basic particle, 2S is a non-negative integer	(5.5)
For some basic particles, $S \geq 3/2$	(5.6)

Core

We start to bridge between physics and IOM

We consider items (5.1) and (5.2) to have significance.
People use the next item to model quantum states of physics particles and systems.

$$\Omega = S(S+1) \quad (5.7)$$

Based on items (4.36), (5.1), (5.2), and (5.7), the next item defines a relevant number of dimensions.

$$D_{*P} = 3 \quad (5.8)$$

We define IOM(#E,#P)

The next item defines the term IOM(#E,#P) for cases in which D_E is odd and D_P is odd. Here and below, the M in IOM can denote methods, math, or models.

IOM(#E,#P) denotes IOM restricted so that the following apply (5.9)

- #E denotes the maximum χ for which $n_{E\chi}$ belongs to an open pair
- #P denotes the maximum χ for which $n_{P\chi}$ belongs to an open pair
- For purposes of computing #E and #P, people can consider that the E1 oscillator and the P1 oscillator constitute an open pair
- If $n_{E\chi}$ belongs to an open pair, each $n_{E\ddot{\iota}}$ for which $\chi \geq \ddot{\iota} \geq 2$ belongs to an open pair
- If $n_{P\chi}$ belongs to an open pair, each $n_{P\ddot{\iota}}$ for which $2 \leq \ddot{\iota} \leq \chi$ belongs to an open pair

We anticipate using concepts related to QI space and IOM

We anticipate using the next items.

Gss.5.1	Non-traditional IOM having D_{*P}=3 correlate with the basic particles and with some properties of basic particles.	(5.10)
Gss.5.2	For basic particles, $\nu=-1$ correlates with basic bosons and their fields. $\nu=-3/2$ correlates with basic fermion particles. $\nu=-1/2$ correlates with fermion fields.	(5.11)
Gss.5.3	For basic particles, $\Omega=+S(S+1)>0$ correlates with QE-like phenomena, $\Omega=0$ correlates with the Higgs boson, and $\Omega=-S(S+1)<0$ correlates with QP-like phenomena.	(5.12)

We summarize and illustrate some findings

The next items provide conditions we think necessary for solutions that correlate with non-zero-mass basic particles. These items correlate with items above.

D_{*P} = 3 pertains (5.13)

The set $S\nu$ = (5.14)
- { −1} for boson fields and particles
- { −1/2} for fermion fields
- { −3/2} for fermion particles

2S is (5.15)
- an even integer for boson particles
- an odd integer for fermion particles

$0 \leq S$ (5.16)
$\Omega = \pm S(S+1)$ (5.17)
$\Omega = \nu(\nu+D-2)$, for $\nu \in S\nu$ and some number D (5.18)
$D \geq 1$ (5.19)

Comments

We suggest research

STR.5.1 To what extent would people benefit from defining and using a value for a D_{*E}?

Section 6 Some physics numbers

Abs.6.1 The mass of a tauon may equal a number computed from 4 physics constants.
Abs.6.2 The mass of a tauon may be $1.776814(\sim 48)\times 10^3$ MeV/c^2.
Abs.6.3 4 physics constants define a series of lengths, including the Planck length.

Context

We note importance of considering numbers

People want physics models to reflect known numbers and predict yet-to-be-measured numbers.

We anticipate computing some physics numbers

In this section, we compute numbers we use elsewhere in this paper.

Core

We relate the mass of a tauon to relative strengths of electromagnetism and gravity

The next item characterizes the relative strengths of electromagnetism and gravity. Here, q_e denotes the charge of an electron, $1/(4\pi\varepsilon_0)$ denotes the Coulomb constant, G_N denotes the gravitational constant, and m_e denotes the mass of an electron. This calculation pertains for electrons and positrons. We base numbers (here and below) on items following item (6.31). The uncertainty-range is approximate.

$$\{(q_e)^2/(4\pi\varepsilon_0)\} \,/\, \{G_N(m_e)^2\} \approx 4.1649(\sim 1)\times 10^{42} \quad (6.1)$$

The next items define β and β'. Here, m_{tauon} denotes the mass of a tauon.

$$(4/3)(\beta^6)^2 = \{(q_e)^2/(4\pi\varepsilon_0)\} / \{G_N(m_e)^2\} \quad (6.2)$$

$$\beta' = m_{tauon} / m_e \quad (6.3)$$

The next items estimate β and β'. For β, we estimate an uncertainty-range based on the uncertainty-range item (6.1) shows. (The uncertainty range item (6.4) shows may be an over-estimate.) For β', we base the uncertainty-range on experimental results.

$$\beta \approx 3.477139(\sim 94)\times 10^3 \quad (6.4)$$

$$\beta' \approx 3.47715(31)\times 10^3 \quad (6.5)$$

The next item pertains.

Gss.6.1 $\beta' = \beta$. (6.6)

We note a series of lengths pertaining to numbers related to electrons and positrons

The next item notes a traditional physics length, the Planck length. We provide a symbol, λ_2, for that length. ħ denotes Planck's constant (reduced). c denotes the speed of light. We add to the traditional statement 2 factors, each of value 1. The first such factor is m_e^0. The second such factor is 2^0.

$$\lambda_2 = G_N^{1/2}\, m_e^0\, \hbar^{1/2}\, c^{-3/2}\, 2^0 \quad (6.7)$$

The next item applies a traditional formula to a property (mass) associated with electrons. The formula represents the Schwarzschild radius. We denote the Schwarzschild radius by R_S. Traditionally, people apply the Schwarzschild-radius formula to black holes. Traditionally, people do not apply the formula to objects people claim have not enough mass to form black holes. We add to the traditional statement 1 factor with value 1. That factor is $\hbar^0$.

$$\lambda_4 = R_S = G_N^1\, m_e^1\, \hbar^0\, c^{-2}\, 2^1 \quad (6.8)$$

The next item shows the ratio of the above 2 lengths.

$$Z = \lambda_2 / \lambda_4 = G_N^{-1/2}\, m_e^{-1}\, \hbar^{1/2}\, c^{1/2}\, 2^{-1} \approx 1.1945\times 10^{22} \quad (6.9)$$

The next item defines a series of lengths.

$$\lambda_\$ = \lambda_2 \cdot Z^{(2-\$)/2} \quad (6.10)$$

The next items show factors and values (for electrons and positrons). These approximate lengths are the products of the factors indicated by the five columns having labels γ for $ύ^γ$ (for some ύ). Times are computed via time=length/c. The time-centric column shows the log-base-10 of times. The time since the big bang is $\sim 10^{17.6}$ seconds. The $ column values indicate possibly interesting correlations between items and e-family bosons we symbolize by $e%&. [Section 8] The $' column values indicate possibly interesting correlations between items and properties of objects. [Items following item (6.23)] In effect, $'=$+2.

$	$\lambda_\$$	Length (m)	Log_{10} (time (sec))	Concept	γ for G_N^γ	γ for m_e^γ	γ for $\hbar^\gamma$	γ for c^γ	γ for 2^γ	$'	(6.11)
		3.3×10^{53}	+45		−1.5	−4	2.5	0.5	−4		(6.12)
		2.7×10^{31}	+23		−1	−3	2	0	−3		(6.13)
		2.3×10^{9}	+0.88		−0.5	−2	1.5	−0.5	−2		(6.14)
0	λ_0	1.9×10^{-13}	−21	spin/mass	0	−1	1	−1	−1	2	(6.15)
2	λ_2	1.6×10^{-35}	−43	Planck length	0.5	0	0.5	−1.5	0	4	(6.16)
4	λ_4	1.4×10^{-57}	−65	R_S	1	1	0	−2	1	6	(6.17)
6	λ_6	1.1×10^{-79}	−87		1.5	2	−0.5	−2.5	2	8	(6.18)
8	λ_8	9.5×10^{-102}	−109.5		2	3	−1	−3	3		(6.19)
		7.9×10^{-124}	−132		2.5	4	−1.5	−3.5	4		(6.20)

The next item pertains.

Gss.6.2 We attach significance to $\lambda_\$$ for which a particle property has an exponent γ=0. (6.21)

G_N is not a particle property. We note item (6.2).

Gss.6.3 Regarding λ_0, people can consider q_e to be a particle property for which $|q_e|^0$ pertains. (6.22)

The next items pertain. The leftmost 3 columns pertain to electrons. The rightmost 2 columns provide names for physical properties.

$'	Factor (for an electron) for which γ=0	Size of the property (for an electron)	Traditional name of the property	Our term for a name for the property	(6.23)
2	q_e	q_e	Charge	PROPE2	(6.24)
4	m_e	m_e	Mass	PROPE4	(6.25)
6	$\hbar$	$(g_S)\hbar/2$	Magnetic moment	PROPE6	(6.26)
8	-	-	-	PROPE8	(6.27)

The next items define candidates for PROPE8 of an object.

PROPE8 denotes fermion count (comprising an object) (6.28)

- Possibly, the object's total number of fermions
- Possibly, the object's net fermion-anti-fermion number - the number of particle fermions (such as electrons) minus the number of anti-particle fermions (such as positrons)
- Possibly, the magnitude of the object's net fermion-anti-fermion number

PROPE8 denotes handedness/chirality (6.29)

- Possibly, the object's overall handedness/chirality
- Possibly, the (gross) sum of the magnitudes of handedness/chirality for basic particles in the object
- Possibly, the net sum of the of handedness/chirality for basic particles in the object
- Possibly the magnitude of the object's net handedness/chirality

PROPE8 denotes an as-yet not specified property of (at least) fermions (6.30)

Possibly, each of PROPE6 and PROPE8 is non-negative. [STR.21.4]

Comments

We show numbers from experiments

The next items show results of experiments. For items (6.44) and (6.45), g_S=2. [Ref.6.1, Ref.6.2, and Ref.6.3]

Symbol	Units	Number	Description	(6.31)
q_e	C	$-1.602176565(35)\times10^{-19}$	charge of an electron	(6.32)
ε_0	$F\ m^{-1}$	$8.854187817\times10^{-12}$	permittivity of free space	(6.33)
G_N	$m^3\ kg^{-1}\ s^{-2}$	$6.67545(18)\times10^{-11}$	gravitational constant	(6.34)
m_e	kg	$9.10938291(40)\times10^{-31}$	mass of an electron	(6.35)
m_e	MeV/c^2	0.510998928(11)	mass of an electron	(6.36)
$\hbar$	J s	$1.054571726(47)\times10^{-34}$	Planck constant, reduced	(6.37)
$\hbar$	MeV s	$6.58211928(15)\times10^{-22}$	Planck constant, reduced	(6.38)
c	$m\ s^{-1}$	2.99792458×10^{8}	speed of light in a vacuum	(6.39)
α	(no units)	$7.2973525698(24)\times10^{-3}$	fine-structure constant	(6.40)
α^{-1}	(no units)	137.035999074(44)		(6.41)
m_{tauon}	MeV/c^2	$1.77682(16)\times10^{3}$	mass of a tauon	(6.42)
m_{muon}	MeV/c^2	105.6583715 ± 0.0000035	mass of a muon	(6.43)
a	(no units)	$(1159.65218076 \pm0.00000027)\times10^{-6}$	electron magnetic moment anomaly $(g - g_S) / g_S$	(6.44)
a	(no units)	$(11659209\pm6)\times10^{-10}$	muon magnetic moment anomaly $(g - g_S) / g_S$	(6.45)

Values for Z, W, and Higgs boson masses appear in Section 17. Values for ranges of quark masses appear in Section 18.

We restate a formula for the mass of a tauon and we estimate that mass

The next item restates results from above.

$$m_{tauon} / m_e = \beta = \exp(\ (1/12)\ \log\{\ \{3/4\}\ \{(q_e)^2/(4\pi\varepsilon_0)\}\ /\ \{G_N(m_e)^2\}\ \}\ ) \quad (6.46)$$

The next item predicts a mass for tauons. We base this prediction on items (6.3), (6.4), and (6.6). The uncertainty range item (6.47) shows may be an over-estimate.

$$m_{tauon} \approx 1.776814(\sim 48)\times 10^3\ \mathrm{MeV}/c^2 \quad (6.47)$$

We note another number

The next items pertain.

$$Z' = (1/4\pi\varepsilon_0)^{-1/2}\,|q_e|^{-1}\,\hbar^{1/2}\,c^{1/2}\,2^{-1} \approx 5.8531 \quad (6.48)$$
$$\alpha = (1/4)\cdot(Z')^{-2} \quad (6.49)$$

We introduce notation we use regarding charge

The next item defines Q', a symbol related to charge. Here, q denotes the charge of an object.

$$Q' = q\ /\ |q_e| \quad (6.50)$$

We note possible significance for the concept of spin/mass length

Possibly the formula for λ_0 pertains to other than electrons and positrons. For Z and W bosons, a spin/mass length may have significance. For pions, a spin/mass length may have significance. The next items pertain.

- A Z-boson spin/mass length is $\sim 2\times 10^{-18}$ meters (6.51)
- People measure spatial dependence for interactions mediated by the weak interaction
- For a separation of $\sim 10^{-18}$ meters between 2 interacting particles, the weak interaction and the electromagnetic interaction have similar magnitudes [Ref.6.4]
- At a separation of $\sim 3\times 10^{-17}$ meters, the weak interaction is less by approximately a factor of 10^4 [Ref.6.4]
- Calculating using a spin/$\hbar$ of 1/2, a pion spin/mass length would be $\sim 0.70\times 10^{-15}$ meters (6.52)
  - Spin/$\hbar$=1/2 pertains to quarks
  - For a pion, spin/$\hbar$=0
  - That length is a factor ~139.6/0.511 or ~273.2 smaller than that for electrons
- An experimental charge radius for charged pions is $0.78^{+0.09}_{-0.10}\times 10^{-15}$ meters [Ref.6.5]

We suggest research

SOR.6.1 Verify (to a smaller than current experimental uncertainty-range) or refute $\beta' = \beta$ and the predicted tauon mass.

STR.6.1 To what extent does a relationship for m_{muon}/m_e involving physical constants exist? (Here we have in mind a parallel to item (6.46) and not item (18.100), the Koide formula.)

We list references

Ref.6.1 T. Quinn et al, Improved Determination of G Using Two Methods, *Phys. Rev. Lett,* 111, 101102, 2013. (http://link.aps.org/doi/10.1103/PhysRevLett.111.101102)

Ref.6.2 J. Beringer et al. (Particle Data Group), *Phys. Rev. D86*, 010001 (2012). (http://pdg.lbl.gov/2012/reviews/rpp2012-rev-phys-constants.pdf)

Ref.6.3 J. Beringer et al. (Particle Data Group), *Phys. Rev. D86*, 010001 (2012). (http://pdg.lbl.gov/2012/tables/rpp2012-sum-leptons.pdf)

Ref.6.4 Particle Data Group, Electroweak (web page), *The Particle Adventure*, Lawrence Berkeley National Laboratory, http://www.particleadventure.org/electroweak.html.

Ref.6.5 G. T. Adylov, et. al., A measurement of the electromagnetic size of the pion from direct elastic pion scattering data at 50 GeV/c, *Nuclear Physics B*, Volume 128, Issue 3, 3 October 1977, pages 461-505. (http://dx.doi.org/10.1016/0550-3213(77)90056-6)

Part 3 Basic bosons and basic fermions

Context

We discuss traditional models for particles

People describe difficulties regarding correlating the Standard Model with nature. For example, people posit that gravitons mediate interactions people associate with gravity. People say the Standard Model does not provide for gravity or gravitons.

We anticipate a new model for particles

We anticipate correlating IOM with all known basic bosons, with gravitons, and with bosons yet to be discovered. We anticipate correlating IOM with all known basic fermions and with possible fermions yet to be discovered.

Core

We preview sections in this part

Section 7 lists families of basic particles.
Section 8 discusses a family of zero-mass basic particles that includes photons and gravitons.
Section 9 discusses families of non-zero-mass basic particles related to and including the Z boson, W bosons, and Higgs boson.
Section 10 discusses a family of particles related to gluons.
Section 11 discusses families of basic particles related to and including leptons and quarks.

Section 7 Families of basic particles

Abs.7.1 A catalog of families of basic particles points to possible yet-to-be-discovered particles.

Context

We note that techniques for cataloging particles exist

Traditional catalogs of basic particles emphasize concepts such as boson or fermion, spin, and charge.

We note that we provide models that correlate with additional particles

We anticipate describing models correlating with possible particles traditionally not catalogued. For example, models correlate with possible fermions for which $S \neq 1/2$. We show, in this section, information that summarizes work in subsequent sections in (this) Part 3.

Core

We show a catalog of types of basic particles

The next items provide bases for cataloging basic particles.

For a basic particle, exactly 1 of the following pertains for S, which denotes spin/ħ and is a non-negative integer (7.1)

- 2S is even
- 2S is odd

For a basic particle, exactly 1 of the following pertains for a parameter Ω (7.2)

- S>0 and Ω = +S(S+1)
- S=0 and Ω = 0
- S>0 and Ω = −S(S+1)

For a basic particle, m denotes mass and exactly 1 of the following pertains (7.3)

- m = 0
- m ≠ 0

Items above correlate with possibilities for 10 families of basic particles. Here 10 = 2 + 2×2×2. The first term (2=1×1×2) corresponds to S=0. The second term (8=2×2×2) corresponds to S≠0. Observations have yet to detect (and models this paper presents below seem to rule out) the 2 possibilities for which 2S is odd and m=0. Observations have yet to detect (and models this paper presents below seem to rule out) the 1 possibility for which S=0 and m=0. We discuss 7 (=10−2−1) families.

The next items define notation for 7 families. Each known basic particle correlates with a family. The examples column lists some known particles. The examples column also lists the (hypothetical) graviton.

2S	Ω	Mass	Traditional theme	Symbol for the family	Examples of basic particles	(7.4)
Even	>0	0		e	Photon, graviton	(7.5)
Even	>0	≠0	Weak	w	Z, W^-, W^+	(7.6)
Even	0	≠0	Higgs	h	Higgs	(7.7)
Even	<0	≠0		o		(7.8)
Even	<0	0	Strong	s	Gluon	(7.9)
Odd	>0	≠0	Leptons	l	Electron, muon	(7.10)
Odd	<0	≠0	Quarks	q	Up, anti-up	(7.11)

We point to yet-to-be-discovered possible basic particles

We predict yet-to-be-discovered possible basic particles correlating with the e-, o-, and q-families.

Comments

We provide perspective regarding families of possible basic particles

The next items pertain to items following item (7.4).

People may consider the e-family to be non-traditional (7.12)

- People may consider that the photon and graviton do not belong in 1 family

People may consider the o-family to be non-traditional (7.13)

- People may consider that no o-family particles have been discovered

Regarding the traditional theme column, ... (7.14)

- People might assign the term electromagnetic to a traditional interpretation of the e-family

We provide perspective regarding IOM representations

For each of some bosons, people can gain insight by using more than 1 IOM model. Models we show in Part 3 minimize each of #E and #P. For example, we use #E=1 for the e- and w-families. And, we use #P=1 for the s- and o-families. In and beyond Part 4, we explore other models.

We discuss spins for basic bosons

The next item summarizes findings that correlate with models we show in Part 3 for basic bosons.

S = maximum((#E−1)/2 , (#P−1)/2) (7.15)

For IOM Part 3 presents, the next items pertain for basic bosons.

Family	#E	#P	S	
				(7.16)
e	1	3, 5, 7, 9	1, 2, 3, 4	(7.17)
s	3	1	1	(7.18)
w	1	3	1	(7.19)
h	1	1	0	(7.20)
o	3, 5, 7, 9, 11	1	1, 2, 3, 4, 5	(7.21)

In and beyond Part 4, we discuss IOM for which the next items pertain for basic bosons.

Family	#E	#P	S	
				(7.22)
e	1, 3, 5, 7	3, 5, 7, 9	1, 2, 3, 4	(7.23)
s	3	3	1	(7.24)
w	3	3	1	(7.25)
h	1	1	0	(7.26)
o	3, 5, 7, 9, 11	3, 5, 7, 9, 11	1, 2, 3, 4, 5	(7.27)

For these IOM, the next item pertains for basic bosons.

S = (#P−1) / 2 (7.28)

We discuss masses for basic bosons

The next items describe 2 possibly distinct concepts regarding mass. Regarding item (7.30), we defer discussion of the possible choice of − in ± until discussion of item (15.5).

G-mass (gravitational mass) (7.29)

- Mass (or, energy/c^2) correlating with gravitation

K-mass (kinematic mass) (7.30)
- Mass (or, energy/c^2) correlating with the m in $E^2 - c^2P^2 = \pm|m^2|c^4$
  - Here, E and P are quantum operators
    - People might correlate E with energy
    - People might correlate P with momentum

Based on item (6.25), we equate PROPE4 with G-mass.
Based on items following item (11.61) and on items following (11.74), the next items pertain.

The m in item (7.3) correlates with K-mass (7.31)
- S=7/2 fermions and S=9/2 fermions would have non-zero K-mass

The m in item (7.3) sometimes might not correlate with G-mass (7.32)
- S=7/2 fermions and S=9/2 fermions would have zero G-mass

Section 8 Photons, gravitons, and so forth (the e-family)

Abs.8.1 The e-family includes photons, gravitons, and 2 other zero-mass basic bosons.
Abs.8.2 Each e-family basic boson has 2 modes (or, polarizations).
Abs.8.3 Each of the 4 e-family basic bosons mediates a force with spatial dependence R^{-2}.
Abs.8.4 The e-family includes coherences of the e-family's 4 basic bosons.
Abs.8.5 E-family coherences provide forces with spatial dependences of R^{-4}, R^{-6}, and R^{-8}.
Abs.8.6 IOM may provide a way to avoid dealing with infinite photon ground-state energy.

Context

We note traditional difficulties regarding having a quantum theory of gravity

Traditionally, people have found difficulties in trying to develop quantum-mechanical theories of gravitation. People have had difficulty unifying electromagnetism and gravity. Here, unifying denotes developing useful quantum theory that encompasses both electromagnetism and gravity. Such unity might, for example, feature a theoretical basis that points to each of the two interactions.

We anticipate unifying electromagnetism and gravity

We develop IOM that correlate with photons and gravitons belonging to a family that includes 4 basic zero-mass bosons. The model correlates with coherent phenomena, such as lasing and entanglement and such as coherences between photons and gravitons.

Core

We model photons and the vector potential

People traditionally describe some aspects of a photon in terms of properties such as energy and momentum. Such properties vary based on the relative motions of observers.

People associate the vector potential with photons. People traditionally describe quantum mechanically the vector potential in terms of math correlating with excitations of 2 harmonic oscillators. This aspect has some invariance regarding observers. (To extent observers can choose differing sets of 2 directions orthogonal to the motion of a photon, this aspect in not completely invariant.)

We use QM-type-DL. The next item shows notation regarding modes (polarizations).

# denotes 0 excitation for an oscillator not participating in a mode (8.1)

The next items describe ground states for each of the 2 modes people attribute to photons. Here, we use IOM(1,3). The representation omits a number for a magnitude of energy or momentum. Such a number depends on an observer that detects the photon. Below, we explain notation denoting modes and particles.

n_{E1}	n_{P1}	n_{P2}	n_{P3}	Mode	(8.2)
0	−1	0	#	2e2	(8.3)
0	−1	#	0	2e3	(8.4)

These descriptions correlate with people's statements that a photon does not excite along its direction of motion. Here, n_{P1} correlates with the direction of motion. Applying an n_{P1}-raising operator to either of these ground states results in a zero amplitude. Applying an n_{P1}-raising operator to any excited state of such a ground state results in a zero amplitude.

The next items describe some excited states for P2 polarization. People might say that the energy is proportional to n_{E1} + 1/2.

n_{E1}	n_{P1}	n_{P2}	n_{P3}	Mode	(8.5)
1	−1	1	#	2e2	(8.6)
2	−1	2	#	2e2	(8.7)
3	−1	3	#	2e2	(8.8)

We discuss equations related to energy, lasing, and entanglements

The next item shows an algebraic relationship. The energy is proportional to n_{E1} + 1/2.

$$n_{E1} = n_{P2} + n_{P3} \quad (8.9)$$

The next items show raising operators. Traditional raising-operator coefficients and lowering-operator coefficients pertain to the P2 and P3 oscillators. Such coefficients correlate with lasing. Traditional coefficients do not necessarily pertain to the E1 oscillator. When both modes are excited, the E1 oscillator correlates with entanglements.

$$a_{P2}^{+} \mid n_{P2} > = (1+n_{P2})^{1/2} \mid n_{P2}+1 > \quad (8.10)$$
$$a_{P3}^{+} \mid n_{P3} > = (1+n_{P3})^{1/2} \mid n_{P3}+1 > \quad (8.11)$$

We discuss QE-like aspects related to photons

In classical physics, people discuss a scalar potential and the vector potential. Per item (8.9), n_{E1} correlates with the scalar potential.

We correlate, for e-family bosons, n_{P1} with spatial dependence of the related force

The next item describes forces intermediated by e-family bosons. For example, for photons, PROPE2 (charge) denotes the property for each object. For photons, ύ=2n_{P1}=−2. [Items following item (6.23)]

Gss.8.1 For e-family members, the force imparted between 2 non-overlapping objects scales as $R^{\acute{\upsilon}}$. Here, $\acute{\upsilon}=2n_{P1}$. Here, R denotes the distance between a center of property of one object and a center of property of the other object. (8.12)

We correlate, for e-family basic bosons, P1 with direction of motion

People might say that the next item applies.

P1 correlates, for e-family basic bosons, with the direction of motion (8.13)

We develop a representation for gravitons

People guess that gravitons provide for the gravitational force. People say that gravitons have S=2, no mass, and 2 polarization modes. People say that gravity correlates with curvature in space time. As far as we know, no one has made a verified detection of a graviton.

Gss.8.2 $D_E=1$, $D_P=5$ solutions provide a model for gravitons. (8.14)

We use item (8.12). The next items model the ground state and first two excited states for P4-polarized gravitons. Here, we use IOM(1,5).

n_{E1}	n_{P1}	n_{P2}	n_{P3}	n_{P4}	n_{P5}	Mode	(8.15)
1	−1	#	#	0	#	4e4	(8.16)
2	−1	#	#	1	#	4e4	(8.17)
3	−1	#	#	2	#	4e4	(8.18)

The next item models the ground state of P5-polarized gravitons.

n_{E1}	n_{P1}	n_{P2}	n_{P3}	n_{P4}	n_{P5}	Mode	(8.19)
1	−1	#	#	#	0	4e5	(8.20)

We contrast models and data for photons and gravitons

The next items contrast models for photons and gravitons.

For photons, the P4-and-P5 oscillator pair is closed (8.21)
For gravitons, the P4-and-P5 oscillator pair is open (8.22)

The next item interprets elements of the left side of item (6.2).

Gss.8.3 In the expression $(4/3)(\beta^6)^2 = \{(q_e)^2/(4\pi\varepsilon_0)\} / \{G_N(m_e)^2\}$, the leftmost exponent 2 represents the number of vertices in a Feynman diagram, β^6 represents the ratio of strengths per channel for electromagnetism and gravity (for an interaction between 2 electrons), 4 represents the number of channels for a photon, and 3 represents the number of channels for a graviton. (8.23)

We base the next item on items (8.21), (8.22), and (8.23).

For photons, the P4-and-P5 oscillator pair correlates with a channel (8.24)

The next item pertains.

Gss.8.4 For photons and gravitons, each of the 3 harmonic oscillator pairs P6-and-P7, P8-and-P9, and P10-and-P11 is closed and correlates with a channel. (8.25)

We discuss notation

The next item pertains.

- For particles in the e-family, we use $ to denote #P−1 (8.26)
- We denote modes and particles by symbols of the form $e...

For example, $=2 for photons and $=4 for gravitons.

We extend the series photons, gravitons, and so forth

We extend a series that begins with photons and gravitons. The next items show ground states for even-polarized zero-mass basic particles. Each item following item (8.27) shows 1 of the 2 modes of an e-family basic boson.

n_{E1}	n_{P1}	n_{P2}	n_{P3}	n_{P4}	n_{P5}	n_{P6}	n_{P7}	n_{P8}	n_{P9}	Mode	(8.27)
0	−1	0	#							2e2	(8.28)
1	−1	#	#	0	#					4e4	(8.29)
2	−1	#	#	#	#	0	#			6e6	(8.30)
3	−1	#	#	#	#	#	#	0	#	8e8	(8.31)

The mode item (8.30) models correlates with 2 channels. The mode item (8.31) models correlates with 1 channel. We do not extend the series to include, for example, a 10e(10) mode. We assume a 10e(10) mode would have no channels. Thus, we assume a 10e(10) mode would not correlate with observable phenomena. Also, we do not extend the series to include, for example, a 0eØ (in which Ø would denote a list with no items). Presumably, for a 0eØ, IOM(1,1) would pertain. For such a 0eØ, no states with $n_{E1}\geq 0$ and $n_{P1}=-1$ would satisfy Œ=0.

The next item pertains.

For e-family basic bosons, 2S = $ = #P−1 (8.32)

We correlate some e-family aspects of IOM with concepts related to traditional physics

People may find that the next item provides insight regarding a mapping of aspects people may associate with e-family IOM into aspects people may associate with 4-dimensional energy-momentum space.

Indices χ for harmonic oscillators n_χ relevant to e-family IOM	Index for an axis in energy-momentum space (for the 1-axis aligned with kinematic momentum)	(8.33)
E1	0	(8.34)
P1	1	(8.35)
P2, P4, P6, and P8	2	(8.36)
P3, P5, P7, and P9	3	(8.37)

People may find that the next item provides insight regarding a mapping of aspects people may associate with e-family IOM into aspects people may associate with space time.

For ï>3, with Pï denoting an oscillator, (8.38)

- Modes having even ï exert influence in the same spatial direction as do modes having ï=2
- Modes having odd ï exert influence in the same spatial direction as do modes having ï=3

We provide notation for e-family members

We denote bosons closely related to the above series by \$e%&. Here, #P=D_P≥3.

\$ = (#P−1)/2 (8.39)

The symbol e denotes related to electromagnetism (8.40)

The list % contains even integers (8.41)

- Values for those integers can be 2, 4, 6, or 8
- Integers appear in ascending order
- The list contains no less than 1 element and no more than 4 elements
- No integer appears more than once
- Each integer ï that appears corresponds to an open Pï-and-P(ï+1) oscillator pair

The symbol & denotes boson (8.42)

- Similar notation without & (and possibly with odd values for ï) describes modes

We list some particles, coherences, modes, and force spatial-dependences

The next items show ground states for the 4 e-family basic bosons. We call these the photon-graviton series bosons.

n_{E1}	n_{P1}	n_{P2}	n_{P3}	n_{P4}	n_{P5}	n_{P6}	n_{P7}	n_{P8}	n_{P9}	Particle	(8.43)
0	−1	0	0							2e2&	(8.44)
1	−1	#	#	0	0					4e4&	(8.45)
2	−1	#	#	#	#	0	0			6e6&	(8.46)
3	−1	#	#	#	#	#	#	0	0	8e8&	(8.47)

The next items show ground states for odd-polarization modes of photon-graviton series bosons.

n_{E1}	n_{P1}	n_{P2}	n_{P3}	n_{P4}	n_{P5}	n_{P6}	n_{P7}	n_{P8}	n_{P9}	Mode	(8.48)
0	−1	#	0							2e3	(8.49)
1	−1	#	#	#	0					4e5	(8.50)
2	−1	#	#	#	#	#	0			6e7	(8.51)
3	−1	#	#	#	#	#	#	#	0	8e9	(8.52)

The next items show some e-family ground states. We call these the maximal-% e-family members. For other than 2e2%, the e-family members shown feature coherences.

n_{E1}	n_{P1}	n_{P2}	n_{P3}	n_{P4}	n_{P5}	n_{P6}	n_{P7}	n_{P8}	n_{P9}	Particle or coherence	(8.53)
0	−1	0	0							2e2&	(8.54)
0	−2	0	0	0	0					4e24&	(8.55)
0	−3	0	0	0	0	0	0			6e246&	(8.56)
0	−4	0	0	0	0	0	0	0	0	8e2468&	(8.57)

The next items show ground states for even-polarization modes of maximal-% e-family members. We use the term coherence to describe modes that encompass more than 1 basic e-family boson. Items (8.60), (8.61), and (8.62) exemplify this concept of coherence.

n_{E1}	n_{P1}	n_{P2}	n_{P3}	n_{P4}	n_{P5}	n_{P6}	n_{P7}	n_{P8}	n_{P9}	Mode	(8.58)
0	−1	0	#							2e2	(8.59)
0	−2	0	#	0	#					4e24	(8.60)
0	−3	0	#	0	#	0	#			6e246	(8.61)
0	−4	0	#	0	#	0	#	0	#	8e2468	(8.62)

The next items show first excited states for maximal-% even-polarization modes.

n_{E1}	n_{P1}	n_{P2}	n_{P3}	n_{P4}	n_{P5}	n_{P6}	n_{P7}	n_{P8}	n_{P9}	Mode	(8.63)
1	−1	1	#							2e2	(8.64)
2	−2	1	#	1	#					4e24	(8.65)
3	−3	1	#	1	#	1	#			6e246	(8.66)
4	−4	1	#	1	#	1	#	1	#	8e2468	(8.67)

The next items show first excited states for some possible mixed-polarization coherence modes. We do not further discuss in this paper the extent to which mixed-polarization coherence modes occur.

n_{E1}	n_{P1}	n_{P2}	n_{P3}	n_{P4}	n_{P5}	n_{P6}	n_{P7}	n_{P8}	n_{P9}	Mode	(8.68)
2	−2	1	#	#	1					4e25	(8.69)
3	−3	1	#	#	1	1	#			6e256	(8.70)
4	−4	1	#	#	1	1	#	#	1	8e2569	(8.71)

The next items show ground states for even-polarization modes for another series of e-family members. Each particle in this series involves excitations of oscillators in the P2-and-P3 oscillator pair. We use the term $e2& series to denote the corresponding e-family members. We do not apply the term coherence to modes for which excitation does not involve at least 2 open P(even)-and-P(odd) pairs.

n_{E1}	n_{P1}	n_{P2}	n_{P3}	n_{P4}	n_{P5}	n_{P6}	n_{P7}	n_{P8}	n_{P9}	Mode	(8.72)
0	−1	0	#							2e2	(8.73)
0	−2	0	#	#	#					4e2	(8.74)
0	−3	0	#	#	#	#	#			6e2	(8.75)
0	−4	0	#	#	#	#	#	#	#	8e2	(8.76)

The next items show spatial dependences for forces correlated with some e-family members.

Particle or coherence	Spatial dependence of force	(8.77)
2e2&	R^{-2}	(8.78)
4e4&	R^{-2}	(8.79)
6e6&	R^{-2}	(8.80)
8e8&	R^{-2}	(8.81)
4e24&	R^{-4}	(8.82)
6e246&	R^{-6}	(8.83)
8e2468&	R^{-8}	(8.84)
4e2&	R^{-2}	(8.85)
6e2&	R^{-2}	(8.86)
8e2&	R^{-2}	(8.87)

We discuss object-properties for which e-family bosons couple basic fermions and other objects

The next items use the series that items following item (6.23) present.

	$e%& forces for which % contains a 2 couple to PROPE2	(8.88)
	$e%& forces for which % contains a 4 couple to PROPE4	(8.89)
Gss.8.5	$e%& forces for which % contains a 6 couple to PROPE6.	(8.90)
Gss.8.6	$e%& forces for which % contains an 8 couple to PROPE8.	(8.91)

Comments

We discuss coherences

When the % in $e%& contains more than 1 element, an excitation provides for coherence between the excitations of the oscillators corresponding to the various elements. The next items show raising operators for 1 mode for each of some coherences. Here, n denotes the quantum number shared by the even-oscillators in open pairs.

Mode	Raising operator	(8.92)
2e2	$a^+ \mid n > = (1+n)^{1/2} \mid n+1 >$	(8.93)
4e24	$a^+ \mid n > = (1+n)^1 \mid n+1 >$	(8.94)
6e246	$a^+ \mid n > = (1+n)^{3/2} \mid n+1 >$	(8.95)
8e2468	$a^+ \mid n > = (1+n)^2 \mid n+1 >$	(8.96)

We discuss spatial dependence of e-family forces

People correlate R^{-2} force behavior for 2e2& with the notion that, in 3-dimensional space, the areas of the surfaces of spheres increase in proportion the square of the radii of the spheres. Items including and following item (8.77) extend such thoughts.

We discuss spin/ħ for e-family members

From work above, the next item pertains for each of the 4 e-family basic bosons. We think this item pertains for all e-family members.

| S = (#P−1)/2 = $/2 (8.97)

We note the topic of the extent to which e-family bosons provide attraction and repulsion

Forces in the e-family can provide attraction or repulsion. In Section 16, we note data that may correlate with 2 maximal-% e-family bosons (other than 2e2&) providing repulsion and 1 maximal-% e-family boson (other than 2e2&) providing attraction.

We note we may have resolved an issue regarding zero-point energy

The next items present the possibility people can consider that this work resolves an issue regarding photon zero-point energy.

| In traditional physics, the sum (over photon states) of ground-state energy is unbounded (8.98)
| IOM feature Œ=0 for bosons (8.99)

We discuss limits on the e-family

Above, we posit a limit of 4 basic bosons for the e-family. Below, we show symmetries that people might correlate with a limit of 4 e-family basic bosons. [Section 13]

We suggest research

SOR.8.1 Detect instances or effects of, or rule out (to some confidence level) the existence of, 4e24& coherences.
SOR.8.2 Determine the extent to which 4e24& includes mixed (even and odd) polarization modes.
SOR.8.3 Conduct experiments to produce or rule out (to some confidence level) reactions that would produce 4e24& from multiply excited 2e2& modes.
SOR.8.4 Measure or infer S for various e-family coherences and various other non-basic-boson e-family members.
SOR.8.5 Determine (to some confidence level) or rule out that each basic e-family boson in a coherence moves in a direction equal to the direction for each other basic e-family boson in the coherence.
SOR.8.6 Measure or infer signs and magnitudes for forces mediated by e-family members other than 2e2& and 4e4&.
STR.8.1 Predict the extent to which e-family coherences exhibit mixed (even and odd) polarization modes.
STR.8.2 Harmonize models and observations or experiments regarding S for e-family members.
STR.8.3 To what extent might people find it appropriate to associate $(t')^0$ behavior with $e%&-mediated interactions? (Here, t' denotes time. [Section 14])

Section 9 Non-zero-mass basic bosons (the w-, h-, and o-families)

Abs.9.1 Families of non-zero-mass basic bosons include the w-family (Z, W^-, and W^+ bosons), the h-family (Higgs boson), and the o-family (for which $\Omega<0$).

Context

We note some known and conjectured non-zero-mass basic bosons

People discuss non-zero-mass basic bosons. The Standard Model correlates with the Z, W^-, W^+, and Higgs bosons. People discuss a concept of leptoquarks.

We anticipate that IOM correlate with yet-to-be-detected non-zero-mass basic bosons

We discuss the o-family.

Core

We discuss IOM correlating with non-zero-mass basic bosons

We use QM-type-CS. Based on items (4.36) and (5.14), the next item pertains for non-zero-mass basic bosons.

D for fields = 3 − Ω (9.1)

Based on known numbers of particles in the w- and h-families, the next item pertains.

Number of particles = 2S+1 (9.2)

We list non-zero-mass basic bosons

We correlate particles with IOM for which ν=−1. The next items pertain. Rows for which the number of particles column shows a blank fall outside assumptions we make. For example, we assume that only solutions having D≥1 pertain. We discuss the limit S≤5 later. [Item (13.40)]

S	Ω	D	Traditional particles	Possible particles	Number of particles	(9.3)
2	6	−3				(9.4)
1	2	1	Z, W^-, W^+		3	(9.5)
0	0	3	Higgs		1	(9.6)
1	−2	5		2o%	3	(9.7)
2	−6	9		4o%	5	(9.8)
3	−12	15		6o%	7	(9.9)
4	−20	23		8o%	9	(9.10)
5	−30	33		10o%	11	(9.11)
...	−S(S+1)					(9.12)

We denote non-zero-mass basic bosons by $ύ%. We use IOM(11,3) and subsets thereof. Here, we do not speculate as to the extent the concept of channels applies and as to the extent $ would correlate with an appropriate number of channels.

$ = 2S (9.13)

These symbols pertain (9.14)

- ύ=w for the w-family
- ύ=h for the h-family
- ύ=o for the o-family

These numbers pertain (9.15)

- $ = 2 for the w-family
- $ = 0 for the h-family
- $ = 2, 4, 6, 8, or 10 for the o-family

% denotes an integer ï (9.16)

- The integer ï correlates with a QI-space harmonic oscillator, ...
  - either Eï or Pï, depending on the family
- For the w-family, oscillators P1, P2, and P3 pertain regarding ï
- For the h-family, ï=1
  - (Each of the E1 and P1 oscillators pertains for the 1 particle, the Higgs boson)
- For the o-family, oscillators E11, E10, ..., E2, and E1 pertain regarding ï

The next items note symbols.

Particle	Symbol	
		(9.17)
Z	2w1	(9.18)
W^-	2w2	(9.19)
W^+	2w3	(9.20)
Higgs	0h1	(9.21)

The next items provide a QM-type-DL model for w-, h-, and o-family ground states. Here, we use EB to denote E11. We use EA to denote E10. In 10o11, the 11 is an eleven, not a 1 followed by a 1. In 10o10, the second 10 is a ten, not a 1 followed by a 0. Here, we confine, for known bosons (Z, W, and Higgs), the use of oscillators Eï to just E1. Here, we confine, for o-family bosons, the use of oscillators Pï to just P1. We use item (7.15) to calculate spin. Later, we show representations for which S=(#P−1)/2. [Section 12]

n_{EB}	n_{EA}	n_{E9}	n_{E8}	n_{E7}	n_{E6}	n_{E5}	n_{E4}	n_{E3}	n_{E2}	n_{E1}	n_{P1}	n_{P2}	n_{P3}	n_{P4}	n_{P5}	Mode	(9.22)
										1	0	#	#			2w1	(9.23)
										1	#	0	#			2w2	(9.24)
										1	#	#	0			2w3	(9.25)
										0	0					0h1	(9.26)
								0	#	#	1					2o3	(9.27)
								#	0	#	1					2o2	(9.28)
								#	#	0	1					2o1	(9.29)
						0	#	#	#	#	2					4o5	(9.30)
						#	0	#	#	#	2					4o4	(9.31)
						#	#	0	#	#	2					4o3	(9.32)
						#	#	#	0	#	2					4o2	(9.33)

n_{EB}	n_{EA}	n_{E9}	n_{E8}	n_{E7}	n_{E6}	n_{E5}	n_{E4}	n_{E3}	n_{E2}	n_{E1}	n_{P1}	n_{P2}	n_{P3}	n_{P4}	n_{P5}	Mode	(9.22)
						#	#	#	#	0	2					4o1	(9.34)
				0	#	#	#	#	#	#	3					6o7	(9.35)
				#	0	#	#	#	#	#	3					6o6	(9.36)
				#	#	0	#	#	#	#	3					6o5	(9.37)
				#	#	#	0	#	#	#	3					6o4	(9.38)
				#	#	#	#	0	#	#	3					6o3	(9.39)
				#	#	#	#	#	0	#	3					6o2	(9.40)
				#	#	#	#	#	#	0	3					6o1	(9.41)
		0	#	#	#	#	#	#	#	#	4					8o9	(9.42)
		#	0	#	#	#	#	#	#	#	4					8o8	(9.43)
		#	#	0	#	#	#	#	#	#	4					8o7	(9.44)
		#	#	#	0	#	#	#	#	#	4					8o6	(9.45)
		#	#	#	#	0	#	#	#	#	4					8o5	(9.46)
		#	#	#	#	#	0	#	#	#	4					8o4	(9.47)
		#	#	#	#	#	#	0	#	#	4					8o3	(9.48)
		#	#	#	#	#	#	#	0	#	4					8o2	(9.49)
		#	#	#	#	#	#	#	#	0	4					8o1	(9.50)
0	#	#	#	#	#	#	#	#	#	#	5					10o11	(9.51)
#	0	#	#	#	#	#	#	#	#	#	5					10o10	(9.52)
#	#	0	#	#	#	#	#	#	#	#	5					10o9	(9.53)
#	#	#	0	#	#	#	#	#	#	#	5					10o8	(9.54)
#	#	#	#	0	#	#	#	#	#	#	5					10o7	(9.55)
#	#	#	#	#	0	#	#	#	#	#	5					10o6	(9.56)
#	#	#	#	#	#	0	#	#	#	#	5					10o5	(9.57)
#	#	#	#	#	#	#	0	#	#	#	5					10o4	(9.58)
#	#	#	#	#	#	#	#	0	#	#	5					10o3	(9.59)
#	#	#	#	#	#	#	#	#	0	#	5					10o2	(9.60)
#	#	#	#	#	#	#	#	#	#	0	5					10o1	(9.61)

Comments

We discuss aspects of the o-family

Work below provides the charges the next items state. [Items (17.20) and (17.21)]

Q' for 2o3 equals +1/3 (9.62)
Q' for 2o2 equals −1/3 (9.63)
Q' for \$o1 equals 0, for \$ = 2, 4, 6, 8, and 10 (9.64)

Work below provides 2 possibilities for the charge of each \$o3 and each \$o2 for which \$ = 4, 6, 8, or 10. [Items (17.25) and (17.27)]

The next items pertain. Wording in these items relies on distinguishing # from 0. A transfer of PROPO1 correlates with a fermion's changing generation. We do not discuss possible interpretations of PROPO10. Here, EB denotes E11. EA denotes E10. The e-family does not interact with PROPO1 or with PROPE10.

Gss.9.1 O-family bosons for which (for the ground state) n_{E1}=0 transfer PROPO1. (9.65)

Gss.9.2	O-family bosons for which (for the ground state) $n_{E3}=0$ or $n_{E2}=0$ transfer PROPE2.	(9.66)
Gss.9.3	O-family bosons for which (for the ground state) $n_{E5}=0$ or $n_{E4}=0$ transfer PROPE4.	(9.67)
Gss.9.4	O-family bosons for which (for the ground state) $n_{E7}=0$ or $n_{E6}=0$ transfer PROPE6.	(9.68)
Gss.9.5	O-family bosons for which (for the ground state) $n_{E9}=0$ or $n_{E8}=0$ transfer PROPE8.	(9.69)
Gss.9.6	O-family bosons for which (for the ground state) $n_{EB}=0$ or $n_{EA}=0$ transfer PROPO10.	(9.70)

We explore such matters further in Section 11, Section 12, and Section 22.

People say that quarks do not exist as free-ranging particles. We find $\Omega<0$ for quarks. [Section 11] We reason by analogy. The next items apply. For item (9.72), we do not state a possibility of pairs, based on work in Section 18.

Gss.9.7	Basic particles for which $\Omega<0$ cannot range freely.	(9.71)
Gss.9.8	O-family basic particles are created in at least triplets.	(9.72)

We discuss fields

Item (4.66) provides for twice as many solutions as we note above in this section. [Item (9.2)] The next items provide an interpretation relevant to each pair `s+1 and `s−1 of solutions.

One linear combination of the 2 solutions correlates with a particle-creation operator	(9.73)
An orthogonal linear combination of the 2 solutions correlates with a particle-destruction operator	(9.74)

We suggest research

SOR.9.1 Verify or rule out (to some confidence level) existence of o-family bosons.

SOR.9.2 Verify or rule out (to some confidence level) that o-family bosons cannot be created singly.

SOR.9.3 Verify or rule out (to some confidence level) changes to nuclear theory people propose based on o-family physics.

STR.9.1 How best might people explore the existence and characteristics of $o% particles?

STR.9.2 What known or new phenomena people might explain based on the o-family?

STR.9.3 To what extent do o-family bosons correspond to aspects of the shell model for atomic nuclei? (Harmonic-oscillator math seems to pertain to each of the o-family and the shell model.)

STR.9.4 To what extent might people explain properties of atomic nuclei, based on o-family bosons (and gluons and other physics)?

STR.9.5 To what extent might people explain properties of neutron stars, based on o-family bosons (and other physics)?

Section 10 Gluons (the s-family)

Abs.10.1 S-family bosons provide for gluons for each of 2 sets of 3 color charges.

Context

We review matters related to the strong interaction

People state that the strong interaction binds quarks into mesons and into baryons. Examples of baryons include protons and neutrons. People state that gluons intermediate the strong interaction. People state that gluons have 0 mass. People state that color charge is a property associated with the strong interaction.

We correlate IOM with gluons

We anticipate describing an s-family and correlating the s-family with gluons.

Core

We list s-family basic bosons

The next items show s-family basic bosons. We use IOM(3,1). Here, #P=1. In items following item (12.16), we show alternative representations in which #P=3 and S=(#P−1)/2=1. For the notation for modes, the letter a correlates with 1 cyclic order of −1, 0, and 1 for QE-like quantum numbers. The letter d correlates with the other cyclic order for QE-like quantum numbers. In the notation 2s(...,ɩ̈"), ɩ̈" denotes the oscillator Eɩ̈" for which $n_{E\ddot{\iota}''}$=1.

n_{E3}	n_{E2}	n_{E1}	n_{P1}	Mode	(10.1)
−1	0	1	1	2s(a,1)	(10.2)
1	−1	0	1	2s(a,3)	(10.3)
0	1	−1	1	2s(a,2)	(10.4)
0	−1	1	1	2s(d,1)	(10.5)
1	0	−1	1	2s(d,3)	(10.6)
−1	1	0	1	2s(d,2)	(10.7)

For each s-family basic boson, the next items pertain. Here, the symbols CCɩ̈ denote color charges. The color charges are CC3, CC2, and CC1 (as in n_{E3}, n_{E2}, and n_{E1}, respectively).

For exactly 1 value of ɩ̈, $n_{E\ddot{\iota}}$=−1 (10.8)
For exactly 1 value of ɩ̈", $n_{E\ddot{\iota}''}$=1 (10.9)
An excitement includes the following (10.10)

- A quark loses color charge CCɩ̈
- $n_{\ddot{\iota}}$ becomes 0
- $n_{\ddot{\iota}''}$ becomes 0

People might say that such an excitement erases color charge CCɩ̈ from a quark.
Similarly, a de-excitement from a state having n_{E3} = n_{E2} = n_{E1} = 0 paints the quark with a color charge.

In the next item, 1 trio consists of items (10.2), (10.3), and (10.4). Here, the cyclic order for quantum numbers is −1, 0, 1. Another trio consists of items (10.5), (10.6), and (10.7). Here, the cyclic order for quantum numbers is −1, 1, 0.

Gss.10.1 One trio of s-family bosons provides for gluons pertaining to quarks people consider to be matter. The other trio pertains to quarks people consider to be antimatter. (10.11)

We relate gluons and s-family basic bosons

For discussion, we assume that the first trio corresponds to quarks (and that the second trio corresponds to anti-quarks). (Possibly, the reversed pairing pertains.) The next item symbolizes a component for a gluon. The right element erases color charge CC2 from a quark. The left element paints color charge CC3.

| item (10.2) > < item (10.3) | (10.12)

People sometimes denote the 3 color charges by r (for red), b (for blue), and g (for green). For such, we use ύ' to denote erasing color charge ύ. We use ó to denote painting color charge ó. For discussion, we assume CC2 corresponds to r and CC3 corresponds to b. The next item restates item (10.12).

br' (10.13)

The next items show 2 gluons of which item (10.13) comprises a component.

$$(rb' + br') / 2^{1/2} \quad (10.14)$$
$$-i(rb' - br') / 2^{1/2} \quad (10.15)$$

The next item provides a way people symbolize another 1 of the 8 gluons.

$$(rr' + bb' - 2gg') / 6^{1/2} \quad (10.16)$$

The next item pertains.

g (in the r, b, g representation) correlates with CC1 (in the CC3, CC2, and CC1 representation) (10.17)

Comments

We note an interpretation

People can consider that $\Omega=-2$ for s-family.

We note the matter of handedness/chirality

We think the totality of the next items correlates with handedness/chirality for quarks (and the opposite handedness for anti-quarks).

A correlating of W^- with a specific 1 (the P2 oscillator) of the P2 oscillator and the P3 oscillator (10.18)

A correlating of even-numbered oscillators (E(even) and P(even)) with each other (and a correlating of odd-numbered oscillators (E(odd) and P(odd)) with each other) (10.19)
A correlating of a specific 1 of the 2 trios of s-family bosons with a specific 1 of quarks and anti-quarks (10.20)

We suggest research

STR.10.1 To what extent might people benefit by considering the possibility that the s-family mediates a force with R^0 spatial dependence? (Here, we have in mind asymptotic freedom.)

Section 11 Basic fermions (the q- and l-families)

Abs.11.1 Families of non-zero-mass basic fermions include the l-family (leptons) and the q-family (for which Ω<0).
Abs.11.2 IOM correlate with leptons, quarks, and related fields.
Abs.11.3 IOM correlate with possible basic fermions with S = 3/2, 7/2, and 9/2.
Abs.11.4 One IOM interpretation correlates with each n-type (or, neutrino-like) basic fermion being its own antiparticle. One IOM interpretation correlates with each basic fermion being distinct from its antiparticle.
Abs.11.5 Each q- or l-family particle is a member of a 3-generation trio.

Context

We discuss people's lack of correlating harmonic oscillators with fermions

As far as we know, people tend to correlate harmonic oscillators with bosons and not with fermions.
As far as we know, people tend to focus on harmonic-oscillator solutions for which 2ν is an even integer. As far as we know, people tend to underutilize solutions for which 2ν is an odd integer.

We discuss a basis for correlating harmonic oscillators with fermions

IOM for which 2ν is an odd integer exist. We anticipate correlating odd-integer solutions with aspects of fermions. We focus on interactions fermions have with bosons.

Core

We define n-type

The next item provides a term for some q- and l-family basic particles. Among S=1/2 basic fermions, n-type correlates with neutrinos.

N-type denotes any q- or l-family basic fermion that (11.1)
- Has zero charge
- Can absorb charge via interactions with a 2w2 (W^- boson), 2w3 (W^+ boson), \$o3 boson, or \$o2 boson

We set a scope for the l- and q-families

We use QM-type-CS. We correlate edge solutions with particles. We correlate inside solutions with fields. Based on items (4.36) and (5.14), the next items pertain for basic fermions.

D for fields = (5 – 4Ω) / 2 (11.2)
D for particles = (21 – 4Ω) / 6 (11.3)

Based on known numbers of particles in the q- and l-families, 1 of the next 2 items pertains. The 2(2S+1) choice correlates with each particle having a distinct antiparticle. We call this choice n-type-S. The D choice correlates with n-type particles being their own antiparticles. We call this choice n-type-D.

n-type-S: (11.4)
- Number of particles per generation = 2(2S+1) for particles

n-type-D: (11.5)
- Number of particles per generation = D for particles

Based on item (4.69), the next item pertains.

Number of particles = 3 × (number of particles per generation) (11.6)

The next items show results. For symbols ἰl and ἰq, ἰ=2S. Rows for which both the traditional particles column and the possible particles column show blanks fall outside assumptions we make. For example, we assume that D≤0 does not model phenomena nature exhibits. Regarding possible cases of non-integer D for particles, for this paper, we assume that a non-integer D does not correlate with particles that nature exhibits. Each # per gen column denotes number of particles per generation. The left one (S\`) of those 2 columns correlates with item (11.4) and with the term n-type-S. The right one (D) of those 2 columns correlates with item (11.5) and with the term n-type-D. We assume that an item for which #-per-gen(S\`) < #-per-gen(D) does not correlate with particles nature exhibits. Thus, we assume S>9/2 does not pertain. [Also, discussion related to items (13.38) and (13.39)]

S	Ω	D for fields	D for particles	Traditional particles	Possible particles	# per gen	# per gen	(11.7)
						S\`	D	(11.8)
3/2	15/4	−5	1					(11.9)
1/2	3/4	1	3	Leptons (1l)		4	3	(11.10)
1/2	−3/4	4	4	Quarks (1q)		4	4	(11.11)
3/2	−15/4	10	6		3q	8	6	(11.12)
5/2	−35/4	20	28/3					(11.13)
7/2	−63/4	34	14		7q	16	14	(11.14)
9/2	−99/4	52	20		9q	20	20	(11.15)
11/2	−143/4	74	82/3					(11.16)
13/2	−195/4	100	36			28	36	(11.17)
…	−S(S+1)							(11.18)

We discuss IOM for l-family particles and for S=1/2 q-family particles

The next items apply for lepton particles and for quark particles. Here, L denotes a principal quantum number for spin-like systems that correlate with $-r_E \le r \le 0$. [Item (4.49)] Here, M denotes a secondary quantum number. We use results from items that start with item (4.65).

For each 1 of 3 generations, the following number of relevant solutions correlates with each of S=1/2 with Ω=+3/4 and S=1/2 with Ω=−3/4 (11.19)

- 2(2S+1) = 4

Gss.11.1 For the l-family, combinations of the 4 solutions correspond to 2 of the 3 possible members of an L=1 set (the M=0 member does not apply) and to the 1 member (M=0) of an L=0 set. (11.20)

Gss.11.2 For the q-family, for S=1/2, combinations of the 4 solutions correspond to 4 of the 5 members of an L=2 set (the M=0 member does not apply). (11.21)

We interpret items following item (11.22) as correlating with and supporting items (11.20) and (11.21). In the next items, M" and M' are integer indices. The items show an orderly array of approximate particle masses for charged leptons and quarks. Here, M' correlates with (but does not necessarily equal) M. For each particle, an item shows $\log_{10}(\text{mass}/m_e)$, charge in units of Q', and particle name.

M"	M'	−3	−2	−1	0	1	...	
								(11.22)
M"								(11.23)
0		0.00 (−1) electron	0.61 (+2/3) up	0.97 (−1/3) down		0.97 (+1/3) anti-down		(11.24)
1			2.26 (−1/3) strange	3.40 (+2/3) charm		3.40 (−2/3) anti-charm		(11.25)
2		2.32 (−1) muon	3.93 (−1/3) bottom	5.51 (+2/3) top		5.51 (−2/3) anti-top		(11.26)
3		3.54 (−1) tauon						(11.27)

The next items correlate item (11.20), item (11.21), and items following item (11.22). Here, we consider all 3 generations.

For each of the rows for which M" = 0, 2, or 3, the following apply (11.28)

- The row includes an instance characterized by L=1
  - M' = −3, 0, and +3 characterize this instance
  - A lepton and anti-lepton pair correlate with M' = −3 and M' = +3
  - No particle correlates with M' = 0 for that row
- A corresponding L = 0, M = 0 row has M" ≤ −3

For each of the rows for which M" = 0, 1, or 2, the following apply (11.29)

- The row includes an instance characterized by L = 2
  - M' = −2, −1, 0, +1, and +2 characterize this instance
  - A quark and anti-quark pair correlate with M' = −2 and M' = +2
  - A quark and anti-quark pair correlate with M' = −1 and M' = +1
  - No particle correlates with M' = 0

For each of the rows for which M" = 0 or 2, the following apply (11.30)

- The row includes an instance characterized by L = 3
  - M' = −3, −2, −1, 0, +1, +2, and +3 characterize this instance
  - A lepton and anti-lepton pair correlate with M' = −3 and M' = +3
  - A quark and anti-quark pair correlate with M' = −2 and M' = +2
  - A quark and anti-quark pair correlate with M' = −1 and M' = +1
  - No particle correlates with M' = 0

The next items pertain for work below.

Gss.11.3 For ï an even positive integer, each of the | $n_{Pï}$, $n_{P(ï+1)}$ > states denoted by | −1 , −F > or by | −F , −1 > corresponds to spin/ħ = 1/2. Here, F is an integer and F≥2. (11.31)

Gss.11.4 For the q-family, for a representation for which there is an even positive integer ï for which the | $n_{Pï}$, $n_{P(ï+1)}$ > state denoted by | −1 , −F > or by | −F , −1 > pertains, there is an even positive integer ó for which the | $n_{E(ó+1)}$, $n_{Eó}$ > state denoted by | −1 , −F > or by | −F , −1 > pertains. (11.32)

The next items show a technique for representing charged leptons (l-family) and quarks (the S=1/2 part of the q-family). We describe each particle's possibility for absorbing a property from an interaction with a w-family, h-family, or 2o% boson. For each of these items, Œ=0. Here, we characterize generation-1 fermions. For each oscillator χ for which we show $n_χ$≤−2, absorption of one or more units of property can occur. If such a χ is E3 or E2, a change in the fermion's charge occurs. (Per work in Section 18, Q'=+1/3 for 2o3, Q'=−1/3 for 2o2, and Q'=0 for 2o1.) If such a χ is E1 or P1, a change of generation occurs. (For generation-1 basic fermions, for χ=E1 or χ=P1, $n_χ$≤−3.) If such a χ is P2, Q'=−1 provides the change in charge. If such a χ is P3, Q'=+1 provides the change in charge.

n_{E9}	n_{E8}	n_{E7}	n_{E6}	n_{E5}	n_{E4}	n_{E3}	n_{E2}	n_{E1}	n_{P1}	n_{P2}	n_{P3}	n_{P4}	n_{P5}	n_{P6}	n_{P7}	n_{P8}	n_{P9}	Particle	(11.33)
								−5	−3	−2	−1							positron	(11.34)
								−5	−3	−1	−2							electron	(11.35)
						−1	−2	−3	−3	−2	−1							up	(11.36)
						−2	−1	−3	−3	−2	−1							anti-down	(11.37)
						−1	−2	−3	−3	−1	−2							down	(11.38)
						−2	−1	−3	−3	−1	−2							anti-up	(11.39)

Items (11.34) and (11.35) correlate with 2 of the 4 l-family solutions. Other items following item (11.33) correlate with 4 of the 4 q-family S=1/2 solutions.

The next items correlate with the other 2 l-family solutions. For each of these items, Œ=0. For n-type-S, each of neutrino-a and neutrino-b correlates with a neutrino. For n-type-D, a linear combination of the items labeled neutrino-a and neutrino-b correlates with a neutrino. Presumably, for n-type-D, the orthogonal linear combination does not represent a particle.

n_{E9}	n_{E8}	n_{E7}	n_{E6}	n_{E5}	n_{E4}	n_{E3}	n_{E2}	n_{E1}	n_{P1}	n_{P2}	n_{P3}	n_{P4}	n_{P5}	n_{P6}	n_{P7}	n_{P8}	n_{P9}	Concept	(11.40)
								−5	−3	−2	−1							neutrino-a	(11.41)
								−5	−3	−1	−2							neutrino-b	(11.42)

We explore IOM for q-family particles with S≥3/2

We interpret items following item (11.7) as correlating with possibilities for q-family basic fermions with S = 3/2, 7/2, and 9/2.

The next items correlate notation with spin.

Symbols	S	(11.43)
1l	1/2	(11.44)
1q	1/2	(11.45)
3q, 3qa, 3qb	3/2	(11.46)
7q, 7qa, 7qb	7/2	(11.47)
9q	9/2	(11.48)

Regarding items below, we note the possibility that F could be a function of the relevant QE-like oscillator pair (E11-and-E10, E9-and-E8, E7-and-E6, E5-and-E4, or E3-and-E2) as well as of the relevant QP-like oscillator pair. We further discuss F in Section 17. [Item (17.65)]

We explore IOM for q-family particles with S=3/2

The next item correlates with S=3/2.

Each q-family basic fermion for which S=3/2 has either $n_{P4}=-1$ and $n_{P5}=-F$ or $n_{P4}=-F$ and $n_{P5}=-1$, for some integer F≥2 (11.49)

Each particle discussed below would have 3 generations.

The next items correlate with the first generations of each of 4 basic particles we associate with an L=2 set. In the notation 3q(ï,ύ), ï correlates with $n_{E\text{ï}}=-F$ and ύ correlates with $n_{P\text{ύ}}=-F$. Here, ï≥4 and ύ≥4.

n_{E9}	n_{E8}	n_{E7}	n_{E6}	n_{E5}	n_{E4}	n_{E3}	n_{E2}	n_{E1}	n_{P1}	n_{P2}	n_{P3}	n_{P4}	n_{P5}	n_{P6}	n_{P7}	n_{P8}	n_{P9}	Particle	(11.50)
				−1	−F	−1	−1	−3	−3	−1	−1	−F	−1					3q(4,4)	(11.51)
				−F	−1	−1	−1	−3	−3	−1	−1	−F	−1					3q(5,4)	(11.52)
				−1	−F	−1	−1	−3	−3	−1	−1	−1	−F					3q(4,5)	(11.53)
				−F	−1	−1	−1	−3	−3	−1	−1	−1	−F					3q(5,5)	(11.54)

Based on $n_\chi=-1$ for χ = E3, E2, P2, and P3, these S=3/2 particles cannot interact with charged o-family bosons or with charged w-family bosons.

The next items correlate with the first generations of S=3/2 n-type particles. For n-type-S, 4 particles exist. For n-type-D, a linear combination of item (11.56) and item (11.57) correlates with 1 particle and a linear combination of item (11.58) and item (11.59) correlates with another particle.

n_{E9}	n_{E8}	n_{E7}	n_{E6}	n_{E5}	n_{E4}	n_{E3}	n_{E2}	n_{E1}	n_{P1}	n_{P2}	n_{P3}	n_{P4}	n_{P5}	n_{P6}	n_{P7}	n_{P8}	n_{P9}	Concept	(11.55)
				−1	−1	−1	−F	−3	−3	−1	−1	−F	−1					3qb(2,4)	(11.56)
				−1	−1	−F	−1	−3	−3	−1	−1	−F	−1					3qa(3,4)	(11.57)
				−1	−1	−1	−F	−3	−3	−1	−1	−1	−F					3qb(2,5)	(11.58)
				−1	−1	−F	−1	−3	−3	−1	−1	−1	−F					3qa(3,5)	(11.59)

For these n-type particles, interactions with o-family charged particles can occur. For these n-type particles, no interactions with W bosons occur.

The next item reviews concepts correlating with S=3/2 fermions.

For S=3/2, for each of 3 generations, the 8=2(2S+1) solutions correlate with an L=2 set and 2 L=0 sets (11.60)

- The L=2 set correlates with 4 particles
  - This set lacks its M=0 member
- Each L=0 set correlates with 2 particles (if n-type-S applies) or correlates with 1 particle (if n-type-D applies)
  - Here, M=0

We explore IOM for q-family particles with S=7/2

The next items correlate with the first generations of 12 basic particles we associate with S=7/2.

n_{E9}	n_{E8}	n_{E7}	n_{E6}	n_{E5}	n_{E4}	n_{E3}	n_{E2}	n_{E1}	n_{P1}	n_{P2}	n_{P3}	n_{P4}	n_{P5}	n_{P6}	n_{P7}	n_{P8}	n_{P9}	Particle	(11.61)
−1	−F	−1	−1	−1	−1	−1	−1	−3	−3	−1	−1	−1	−1	−1	−1	−F	−1	7q(8,8)	(11.62)
−F	−1	−1	−1	−1	−1	−1	−1	−3	−3	−1	−1	−1	−1	−1	−1	−F	−1	7q(9,8)	(11.63)
−1	−F	−1	−1	−1	−1	−1	−1	−3	−3	−1	−1	−1	−1	−1	−1	−1	−F	7q(8,9)	(11.64)
−F	−1	−1	−1	−1	−1	−1	−1	−3	−3	−1	−1	−1	−1	−1	−1	−1	−F	7q(9,9)	(11.65)
−1	−1	−1	−F	−1	−1	−1	−1	−3	−3	−1	−1	−1	−1	−1	−1	−F	−1	7q(6,8)	(11.66)
−1	−1	−F	−1	−1	−1	−1	−1	−3	−3	−1	−1	−1	−1	−1	−1	−F	−1	7q(7,8)	(11.67)
−1	−1	−1	−F	−1	−1	−1	−1	−3	−3	−1	−1	−1	−1	−1	−1	−1	−F	7q(6,9)	(11.68)
−1	−1	−F	−1	−1	−1	−1	−1	−3	−3	−1	−1	−1	−1	−1	−1	−1	−F	7q(7,9)	(11.69)
−1	−1	−1	−1	−1	−F	−1	−1	−3	−3	−1	−1	−1	−1	−1	−1	−F	−1	7q(4,8)	(11.70)
−1	−1	−1	−1	−F	−1	−1	−1	−3	−3	−1	−1	−1	−1	−1	−1	−F	−1	7q(5,8)	(11.71)
−1	−1	−1	−1	−1	−F	−1	−1	−3	−3	−1	−1	−1	−1	−1	−1	−1	−F	7q(4,9)	(11.72)
−1	−1	−1	−1	−F	−1	−1	−1	−3	−3	−1	−1	−1	−1	−1	−1	−1	−F	7q(5,9)	(11.73)

The next items correlate with first generations of S=7/2 n-type particles. For n-type-S, 4 particles exist. For n-type-D, a linear combination of item (11.75) and item (11.76) correlates with 1 particle and a linear combination of item (11.77) and item (11.78) correlates with another particle.

n_{E9}	n_{E8}	n_{E7}	n_{E6}	n_{E5}	n_{E4}	n_{E3}	n_{E2}	n_{E1}	n_{P1}	n_{P2}	n_{P3}	n_{P4}	n_{P5}	n_{P6}	n_{P7}	n_{P8}	n_{P9}	Concept	(11.74)
−1	−1	−1	−1	−1	−1	−1	−F	−3	−3	−1	−1	−1	−1	−1	−1	−F	−1	7qb(2,8)	(11.75)
−1	−1	−1	−1	−1	−1	−F	−1	−3	−3	−1	−1	−1	−1	−1	−1	−F	−1	7qa(3,8)	(11.76)
−1	−1	−1	−1	−1	−1	−1	−F	−3	−3	−1	−1	−1	−1	−1	−1	−1	−F	7qb(2,9)	(11.77)
−1	−1	−1	−1	−1	−1	−F	−1	−3	−3	−1	−1	−1	−1	−1	−1	−1	−F	7qa(3,9)	(11.78)

For these n-type particles, interactions with o-family charged particles can occur. For these n-type particles, no interactions with W bosons occur.

The next item reviews concepts correlating with S=7/2 fermions.

For S=7/2, for each of 3 generations, the 16=2(2S+1) solutions correlate with 3 L=2 sets and 2 L=0 sets (11.79)

- Each L=2 set correlates with 4 particles
  - Each such set lacks its M=0 member
- Each L=0 set correlates with 2 particles (if n-type-S applies) or correlates with 1 particle (if n-type-D applies)
  - Here, M=0

We explore IOM for q-family particles with S=9/2

The next items correlate with the first generations of 20 basic particles we associate with S=9/2. Here, EB denotes E11. EA denotes E10. Here, we show symbolically that $n_{P3} = n_{P4} = n_{P5} = n_{P6} = -1$ for each 9q basic particle.

n_{EB}	n_{EA}	n_{E9}	n_{E8}	n_{E7}	n_{E6}	n_{E5}	n_{E4}	n_{E3}	n_{E2}	n_{E1}	n_{P1}	n_{P2}	…	…	n_{P7}	n_{P8}	n_{P9}	Particle	(11.80)
−1	−F	−1	−1	−1	−1	−1	−1	−1	−1	−3	−3	−1	−1	−1	−1	−F	−1	9q(10,8)	(11.81)
−F	−1	−1	−1	−1	−1	−1	−1	−1	−1	−3	−3	−1	−1	−1	−1	−F	−1	9q(11,8)	(11.82)
−1	−F	−1	−1	−1	−1	−1	−1	−1	−1	−3	−3	−1	−1	−1	−1	−1	−F	9q(10,9)	(11.83)
−F	−1	−1	−1	−1	−1	−1	−1	−1	−1	−3	−3	−1	−1	−1	−1	−1	−F	9q(11,9)	(11.84)
−1	−1	−1	−F	−1	−1	−1	−1	−1	−1	−3	−3	−1	−1	−1	−1	−F	−1	9q(8,8)	(11.85)
−1	−1	−F	−1	−1	−1	−1	−1	−1	−1	−3	−3	−1	−1	−1	−1	−F	−1	9q(9,8)	(11.86)
−1	−1	−1	−F	−1	−1	−1	−1	−1	−1	−3	−3	−1	−1	−1	−1	−1	−F	9q(8,9)	(11.87)
−1	−1	−F	−1	−1	−1	−1	−1	−1	−1	−3	−3	−1	−1	−1	−1	−1	−F	9q(9,9)	(11.88)
−1	−1	−1	−1	−1	−F	−1	−1	−1	−1	−3	−3	−1	−1	−1	−1	−F	−1	9q(6,8)	(11.89)
−1	−1	−1	−1	−F	−1	−1	−1	−1	−1	−3	−3	−1	−1	−1	−1	−F	−1	9q(7,8)	(11.90)
−1	−1	−1	−1	−1	−F	−1	−1	−1	−1	−3	−3	−1	−1	−1	−1	−1	−F	9q(6,9)	(11.91)
−1	−1	−1	−1	−F	−1	−1	−1	−1	−1	−3	−3	−1	−1	−1	−1	−1	−F	9q(7,9)	(11.92)
−1	−1	−1	−1	−1	−1	−1	−F	−1	−1	−3	−3	−1	−1	−1	−1	−F	−1	9q(4,8)	(11.93)
−1	−1	−1	−1	−1	−1	−F	−1	−1	−1	−3	−3	−1	−1	−1	−1	−F	−1	9q(5,8)	(11.94)
−1	−1	−1	−1	−1	−1	−1	−F	−1	−1	−3	−3	−1	−1	−1	−1	−1	−F	9q(4,9)	(11.95)
−1	−1	−1	−1	−1	−1	−F	−1	−1	−1	−3	−3	−1	−1	−1	−1	−1	−F	9q(5,9)	(11.96)
−1	−1	−1	−1	−1	−1	−1	−1	−1	−F	−3	−3	−1	−1	−1	−1	−F	−1	9q(2,8)	(11.97)
−1	−1	−1	−1	−1	−1	−1	−1	−F	−1	−3	−3	−1	−1	−1	−1	−F	−1	9q(3,8)	(11.98)
−1	−1	−1	−1	−1	−1	−1	−1	−1	−F	−3	−3	−1	−1	−1	−1	−1	−F	9q(2,9)	(11.99)
−1	−1	−1	−1	−1	−1	−1	−1	−F	−1	−3	−3	−1	−1	−1	−1	−1	−F	9q(3,9)	(11.100)

For S=9/2, no n-type particles exist.
The next item reviews concepts correlating with S=9/2 fermions.

For S=9/2, for each of 3 generations, the 20=2(2S+1) solutions correlate with 5 L=2 sets (11.101)

- Each L=2 set correlates with 4 particles
  - Each such set lacks its M=0 member

Comments

We discuss properties of representations for basic fermions

The following items pertain to representations above.

For the q-family (11.102)
- For generation-1 basic fermions, n_{E1}=−3 and n_{P1}=−3

For the l-family (11.103)
- For generation-1 basic fermions, n_{E1}=−5 and n_{P1}=−3

For the S=1/2 part of the q-family (11.104)
- In any 1 q-family column in the table that starts with item (11.22), the charge for M"=0 differs from the charge for M"=1 or M"=2 (and the last 2 charges are equal)

For the l-family (all of which has S=1/2) (11.105)
- In any 1 l-family column in the table that starts with item (11.22), the charge for M"=0 is the same as the charge for M"=2 or M"=3
- The table that starts with item (11.22) does not list neutrinos

For the l-family (all of which has S=1/2) (11.106)
- Neutrinos do not fall in the range 0≤M"≤3

For the l-family (11.107)
- Representations above for positron and for neutrino-a are similar
- Representations above for electron and for neutrino-b are similar

For the l-family and for the S=1/2 members of the q-family (11.108)
- For generation-1
  - $-6 = n_{P1}+n_{P2}+n_{P3}$
  - $(n_{P1}+n_{P2}+n_{P3})/3$ is a negative integer
- For generation-2 and generation-3
  - $-5 \leq n_{P1}+n_{P2}+n_{P3} \leq -4$
  - $(n_{P1}+n_{P2}+n_{P3})/3$ is a negative non-integer

We discuss interactions between q-family basic fermions and e-family basic bosons

We think that statements above [Section 11] about lack of interactions between basic fermions and w-family, h-family, and 2o% bosons carry over to statements about lack of interactions between q-family basic fermions and e-family basic bosons. The relevant statements feature values of −1 in each of a Pï column and a P(ï+1) column, for which ï is an even integer. If so, IOM predict the next items. We use the term directly to exclude indirect interactions via creation of virtual pairs of particles.

Q-family basic fermions	Direct interaction	No direct interaction	(11.109)
S=1/2	2e2&, 4e4&, 6e6&, 8e8&	-	(11.110)
S=3/2	4e4&, 6e6&, 8e8&	2e2&	(11.111)
S=7/2	8e8&	2e2&, 4e4&, 6e6&	(11.112)
S=9/2	-	2e2&, 4e4&, 6e6&, 8e8&	(11.113)

The next item pertains.

For S≥7/2 basic fermions ... (11.114)

- G-mass = 0
- K-mass ≠ 0

We discuss fermion fields

We note items (4.67) and (4.69). For a given combination of S and Ω, as many field-oriented solution sets exist as do particle-oriented solution sets. The number of field-oriented solutions is 2(2S+1). For particles that correlate with L≠0 sets, pairing a particle and an antiparticle pairs 2 field solutions. We assume 1 linear combination of the 2 field solutions correlates with creating a particle-antiparticle pair. The orthogonal linear combination of the 2 field solutions correlates with destroying a particle-antiparticle pair. For n-type-S, similar considerations apply. For n-type-D, similar considerations possibly apply to the components for particles that correlate with L=0 sets.

The next items pertain, based on comparing the number of field-correlated solutions and the number of particle-correlated solutions.

Fermion fields do not correlate with generation (11.115)
Fermion pair creation (or pair annihilation) that correlates with fields does not correlate with generation (11.116)

We review matters regarding spins for basic fermions

The next items pertain for interaction-based representations of basic fermions.

D_{*P} = 3 pertains (11.117)
S = (#P−2)/2 (11.118)

- Here, #P denotes the minimum value of #P for which IOM(#E,#P) describes the particle

We suggest research

SOR.11.1 Detect (or infer) or rule out (to some confidence level) the existence of basic fermions for which S = 3/2, 7/2, or 9/2.
SOR.11.2 Rule out (to some confidence level) or detect the existence of basic fermions for which S = 5/2 or 11/2.
SOR.11.3 Measure or infer properties of S≥3/2 basic fermions.
SOR.11.4 Measure or infer reactions in which S≥3/2 basic fermions participate.
SOR.11.5 Verify or rule out (to some confidence level) q- and l-family interaction rules we show regarding the w-, h-, and o-families. Determine strengths for interactions for which strengths are yet to be determined.
SOR.11.6 To what extent does either n-type model (n-type-S or n-type-D) better correlate with observations than does the other n-type model?
STR.11.1 Estimate properties PROPE2, PROPE4, PROPE6, PROPE8, and PROPO10 for basic fermions for which S≥3/2.
STR.11.2 Estimate interaction strengths for interactions between w-, h-, and o-family bosons and S≥3/2 basic fermions.
STR.11.3 Describe possible composite objects (such as nuclei or atoms) for which S≥3/2 basic fermions would be components.

STR.11.4 To what extent might people benefit by considering that generation-2 and generation-3 fermions might not be truly basic particles (and are excitations of generation-1 basic particles)?

Part 4 Symmetries and kinematics

Context

We discuss traditional symmetries and equations of motion

People correlate symmetries, such as SU(3)×SU(2)×U(1), with the Standard Model for elementary particles. People discuss possibilities, such as SU(5), for larger symmetries.

People develop quantum mechanical analogs to classical-mechanical equations of motion.

We anticipate finding and extending some traditional results

We anticipate finding and extending Standard Model symmetries.

We anticipate developing directly quantum-mechanical kinematic equations.

Core

We preview sections in this part

Section 12 shows that IOM correlate with the Standard Model symmetry SU(3)×SU(2)×U(1).

Section 13 shows symmetries beyond SU(3)×SU(2)×U(1) possibly pertaining to basic particles.

Section 14 derives $E^2-c^2P^2=0$ for the e- and s-families.

Section 15 derives relationships between E^2 and P^2 for the other 5 families of basic particles.

Section 12 IOM representations and Standard Model symmetries

Abs.12.1 IOM representations correlate with SU(3)×SU(2)×U(1) (or, Standard Model) symmetry for the strong, weak, and electromagnetic interactions.

Abs.12.2 IOM representations can correlate with the statement S=(#P−1)/2 for basic bosons.

Context

We note some aspects of traditional physics

When modeling 1 system, people can use multiple CO-type-ï coordinate systems. For example, for modeling phenomena influenced by a zero-charge non-rotating black hole, people can use Kruskal-Szekeres coordinates or Schwarzschild coordinates.

People use the Standard Model to characterize properties of and interactions between known particles - basic and compound.

We anticipate using multiple representations and finding Standard Model symmetries

We explore correlating multiple IOM representations with physics.

We anticipate showing Standard Model symmetries.

Core

We review values of n_{E1} for some representations of e-family ground states

The next items summarize some concepts related to e-family basic bosons and to maximal-% e-family bosons. Above, we use IOM for which #E=1. [Section 8]

Boson	Basic boson	Maximal-% boson	#E	#P	n_{E1} for the ground state	(12.1)
2e2&	yes	yes	1	3	0	(12.2)
4e4&	yes	no	1	5	1	(12.3)
6e6&	yes	no	1	7	2	(12.4)
8e8&	yes	no	1	9	3	(12.5)
4e24&	no	yes	1	5	0	(12.6)
6e246&	no	yes	1	7	0	(12.7)
8e2468&	no	yes	1	9	0	(12.8)

We explore alternative IOM representations for the 3 e-family basic bosons other than 2e2%

The next items show alternative IOM representations for ground states of e-family basic bosons. Here, Œ=0. Here, n_{E1}=0. Here, #E = 3, 5, and 7 for, respectively, 4e4&, 6e6&, and 8e8&.

n_{E9}	n_{E8}	n_{E7}	n_{E6}	n_{E5}	n_{E4}	n_{E3}	n_{E2}	n_{E1}	n_{P1}	n_{P2}	n_{P3}	n_{P4}	n_{P5}	n_{P6}	n_{P7}	n_{P8}	n_{P9}	Boson	(12.9)
								0	−1	0	0							2e2&	(12.10)
						0	0	0	−1	#	#	0	0					4e4&	(12.11)
				0	0	0	0	0	−1	#	#	#	#	0	0			6e6&	(12.12)
		0	0	0	0	0	0	0	−1	#	#	#	#	#	#	0	0	8e8&	(12.13)

We review symmetries related to the Standard Model

People say that the next item characterizes symmetries associated with the Standard Model.

$$SU(3)\times SU(2)\times U(1) \quad (12.14)$$

We use the term SMS as an acronym for the above Standard Model symmetry.

$$SMS = SU(3)\times SU(2)\times U(1) \quad (12.15)$$

We show and use an alternative representation for s-family basic bosons

The next items provide alternative representations for s-family basic bosons. Here, #P=3 and S=(#P−1)/2.

n_{E3}	n_{E2}	n_{E1}	n_{P1}	n_{P2}	n_{P3}	n_{P4}	n_{P5}	n_{P6}	n_{P7}	n_{P8}	n_{P9}	Mode	(12.16)
−1	0	1	0	0	0							2s(a,1)	(12.17)
1	−1	0	0	0	0							2s(a,3)	(12.18)
0	1	−1	0	0	0							2s(a,2)	(12.19)
0	−1	1	0	0	0							2s(d,1)	(12.20)
1	0	−1	0	0	0							2s(d,3)	(12.21)
−1	1	0	0	0	0							2s(d,2)	(12.22)

People might say that QE-like aspects of items following item (12.16) exhibit an SU(2)×U(1) symmetry. The U(1) component correlates with a choice between 2 cyclic orders. The SU(2) component correlates with 3 items for each cyclic order. People might say that QP-like aspects of items following item (12.16) exhibit an SU(3) symmetry.

The next items pertain. Item (12.24) correlates with the concept that, for each s-family member, at least 1 of n_{E3} and n_{E2} is relevant and, for each l-family member, each of n_{E3} and n_{E2} is not relevant.

The s-family exhibits, with respect to quarks, SMS symmetry	(12.23)
The s-family does not interact with leptons	(12.24)

We explore and use alternative representations for w-, h-, and o-family basic bosons

The next items provide possible alternative representations for w, h, and o-family basic bosons. (For brevity, we do not show details for 10o% basic bosons. Also, we do not show columns for oscillators E11, E10, P10, and P11.) Here, S = (#P−1)/2.

n_{E9}	n_{E8}	n_{E7}	n_{E6}	n_{E5}	n_{E4}	n_{E3}	n_{E2}	n_{E1}	n_{P1}	n_{P2}	n_{P3}	n_{P4}	n_{P5}	n_{P6}	n_{P7}	n_{P8}	n_{P9}	Mode	(12.25)
						0	0	0	0	#	#							2w1	(12.26)
						0	0	0	#	0	#							2w2	(12.27)
						0	0	0	#	#	0							2w3	(12.28)
								0	0									0h1	(12.29)
						0	#	#	0	0	0							2o3	(12.30)
						#	0	#	0	0	0							2o2	(12.31)
						#	#	0	0	0	0							2o1	(12.32)
				0	#	#	#	#	0	0	0	0	0					4o5	(12.33)
				#	0	#	#	#	0	0	0	0	0					4o4	(12.34)
				#	#	0	#	#	0	0	0	0	0					4o3	(12.35)
				#	#	#	0	#	0	0	0	0	0					4o2	(12.36)
				#	#	#	#	0	0	0	0	0	0					4o1	(12.37)
		0	#	#	#	#	#	#	0	0	0	0	0	0	0			6o7	(12.38)
		#	0	#	#	#	#	#	0	0	0	0	0	0	0			6o6	(12.39)
		#	#	0	#	#	#	#	0	0	0	0	0	0	0			6o5	(12.40)
		#	#	#	0	#	#	#	0	0	0	0	0	0	0			6o4	(12.41)
		#	#	#	#	0	#	#	0	0	0	0	0	0	0			6o3	(12.42)
		#	#	#	#	#	0	#	0	0	0	0	0	0	0			6o2	(12.43)
		#	#	#	#	#	#	0	0	0	0	0	0	0	0			6o1	(12.44)
0	#	#	#	#	#	#	#	#	0	0	0	0	0	0	0	0	0	8o9	(12.45)
#	0	#	#	#	#	#	#	#	0	0	0	0	0	0	0	0	0	8o8	(12.46)
#	#	0	#	#	#	#	#	#	0	0	0	0	0	0	0	0	0	8o7	(12.47)
#	#	#	0	#	#	#	#	#	0	0	0	0	0	0	0	0	0	8o6	(12.48)
#	#	#	#	0	#	#	#	#	0	0	0	0	0	0	0	0	0	8o5	(12.49)

n_{E9}	n_{E8}	n_{E7}	n_{E6}	n_{E5}	n_{E4}	n_{E3}	n_{E2}	n_{E1}	n_{P1}	n_{P2}	n_{P3}	n_{P4}	n_{P5}	n_{P6}	n_{P7}	n_{P8}	n_{P9}	Mode	(12.25)
#	#	#	#	#	0	#	#	#	0	0	0	0	0	0	0	0	0	8o4	(12.50)
#	#	#	#	#	#	0	#	#	0	0	0	0	0	0	0	0	0	8o3	(12.51)
#	#	#	#	#	#	#	0	#	0	0	0	0	0	0	0	0	0	8o2	(12.52)
#	#	#	#	#	#	#	#	0	0	0	0	0	0	0	0	0	0	8o1	(12.53)
…																	…	10o%	(12.54)

An SU(3) symmetry pertains to QE-like aspects of the w-family bosons (2w%). To the extent S=1/2 fermions differentiate from each other via handedness, an SU(2)×U(1) symmetry pertains to QP-like aspects of the w-family. Here, SU(2) correlates with P2 and P3. Here, U(1) correlates with fermion handedness.

The next items pertain.

The w-family exhibits, with respect to quarks, SMS symmetry (12.55)

The w-family exhibits, with respect to charged leptons, SMS symmetry (12.56)

- To the extent n-type-S pertains to neutrinos, the w-family exhibits, with respect to neutrinos, SMS symmetry

We discuss 2e2& (photons) and SMS

For e-family basic bosons, S=(#P−1)/2. [Section 8]

For photons (2e2&), an SU(2) symmetry correlates with P2 and P3, just as for the w-family. Also, fermions provide for a U(1) symmetry, just as for the w-family. For e-family basic bosons, QI space features aligning P1 with a direction of motion. Each of energy-momentum space and space time does not feature such a preferred direction. We ascribe an SU(3) symmetry to a relationship between QI space and either of energy-momentum space and space time.

The next items pertain.

Photons (2e2&) exhibit, with respect to quarks, SMS symmetry (12.57)

Photons (2e2&) exhibit, with respect to charged leptons, SMS symmetry (12.58)

- To the extent n-type-S pertains to neutrinos, photons (2e2&) exhibit, with respect to neutrinos, SMS symmetry

We correlate IOM, SMS, and Standard Model particles

The next item follows from items (12.23), (12.24), (12.55), (12.56), (12.57), and (12.58).

IOM exhibit SMS for the combination of the Standard Model fermions (quarks and leptons, except possibly n-type-D neutrinos) and the Standard Model interactions (strong, electromagnetic, and weak) (12.59)

Comments

We note that, for each basic boson, a representation exists in which S=(#P−1)/2

For each basic boson, we show (or allude to) above in this Section 12 an IOM representation for which S=(#P−1)/2.

For s-, w-, h-, and o-family basic bosons, for the respective representations, the next item pertains.

$$\#E = \#P \tag{12.60}$$

For e-family basic bosons, for the respective representations, the next item pertains.

| #E = #P – 2 (12.61)

Section 13 Symmetries beyond the Standard Model

Abs.13.1 SU(7) provides a relevant extension of Standard Model symmetry.

Context

We note some aspects of traditional physics

People use the Standard Model to characterize properties of and interactions between known elementary particles - basic and compound. People recognize limits to the applicability of extant models. People try to find larger symmetries that apply.

We anticipate discovering relevant symmetries

We anticipate providing a non-traditional candidate for a symmetry encompassing Standard Model symmetry.

Core

We correlate IOM, SMS, and a subset of the o-family

With respect to interactions with quarks, IOM provide that the 2o% basic bosons have the same symmetry as do the 2w% particles. With respect to interactions with leptons, 2o% have the same symmetry as do s-family particles. The next item pertains.

| IOM exhibit SMS for the combination of the Standard Model fermions (quarks and leptons, except possibly n-type-D neutrinos) and interactions mediated by 2o% basic bosons (13.1)

We explore QE-like symmetries correlating with the e-family

The next item pertains. Here, each symmetry pertains to the relevant set of QE-like oscillators. For (ï+1)e(ï+1)&, the SIDE set has ï members. The members of the set are Eï through E1. For a ground state, $n_{E\ddot{\imath}}$=0 for each such Eï. [Item (4.13) and items following item (12.9)]

| Gss.13.1 Symmetries related to SU(3), SU(5), and SU(7) pertain for, respectively, 4e4&, 6e6&, and 8e8&. (13.2)

The next item shows the number of generators associated with each SU(ύ).

| The number of generators for SU(ύ) is $ύ^2–1$ (13.3)

The next items show numbers of generators for some values of ύ. The term multiple of ó refers to multiple of ó number of generators. We list only integer multiples.

ύ	Number of generators	Multiple of 8	Multiple of 24	Multiple of 48	
					(13.4)
3	8	1	-	-	(13.5)
5	24	3	1	-	(13.6)
7	48	6	2	1	(13.7)
9	80	10	-	-	(13.8)
11	120	15	5	-	(13.9)
13	168	21	7	-	(13.10)
15	224	28	-	-	(13.11)
17	288	36	12	6	(13.12)

80 is not an integer multiple of 48. We think some applicability of these symmetries ends with item (13.7).

Seemingly, we identify 2 facets pointing toward a limit of S≤4 for the e-family. These facets include the 4-3-2-1 series of channels [Remarks following item (8.31)] and the notion that SU(9) does not apply [Item (13.8)].

The next items apply.

Regarding the e-family, #P≤9 applies for all IOM representations (13.13)
Regarding the e-family, #E≤7 applies for all IOM representations (13.14)

We discuss possibilities for a 48-fold symmetry

The next item shows the product of the respective numbers of generators for the SMS.

$$48 = 8 \times 3 \times 2 \quad (13.15)$$

This 48 for SMS equals the 48 that would pertain to a possible 48-fold symmetry correlating with item (13.7).

We note that people seek symmetries that include SMS

People suggest various groups (including SU(5)) that may both be relevant regarding physics and include mathematically the product of groups that item (12.14) shows. [Ref.13.1]

We note instances of symmetries beyond SMS

The next items pertain. Here, we use fermion IOM from Section 11 and boson IOM from Section 12.

Family	Type of symmetry	S	Symmetry	(13.16)
e	QE-like	2	SU(3)	(13.17)
e	QE-like	3	SU(5)	(13.18)
e	QE-like	4	SU(7)	(13.19)
q	QP-like	3/2	SU(3)	(13.20)
q	QP-like	7/2	SU(7)	(13.21)
q	QP-like	9/2	SU(9)	(13.22)
o	QP-like	2	SU(5)	(13.23)
o	QP-like	3	SU(7)	(13.24)
o	QP-like	4	SU(9)	(13.25)
o	QP-like	5	SU(11)	(13.26)

We interpret correlations with various groups

The next item pertains.

SU(7) is a larger group for SMS (13.27)

Comments

We discuss some group-theory math

The next item shows a result from group theory.

For integers p>1 and n−p>1, (13.28)
- SU(n) ⊃ SU(p)×SU(n−p)×U(1)

The next items provide examples. In a product, such as the right-hand side of either example shows, the order of the groups does not matter.

SU(7) ⊃ SU(2)×U(1)×SU(5) (13.29)
SU(5) ⊃ SU(3)×SU(2)×U(1) (13.30)

We discuss other possibly relevant symmetries

The next items pertain.

(1 + 2/1) (generators for SU(3)) = (3/1) 8 = 24 = generators for SU(5) (13.31)
(1 + 1/1) (generators for SU(5)) = (2/1) 24 = 48 = generators for SU(7) (13.32)
(1 + 2/3) (generators for SU(7)) = (5/3) 48 = 80 = generators for SU(9) (13.33)
(1 + 1/2) (generators for SU(9)) = (3/2) 80 = 120 = generators for SU(11) (13.34)
(1 + 2/5) (generators for SU(11)) = (7/5) 120 = 168 = generators for SU(13) (13.35)
(1 + 1/3) (generators for SU(13)) = (4/3) 168 = 224 = generators for SU(15) (13.36)
(1 + 2/7) (generators for SU(15)) = (9/7) 224 = 288 = generators for SU(17) (13.37)

Possibly, the 2/3 in item (13.33) correlates with 2 out of 3 of the following families - o, q, and e. Above, we show that S=4 and SU(7) correlate with limits regarding the e-family. [Discussion before item (13.13)] 2 families - o and q - remain for symmetries larger than SU(7). The next item pertains.

| SU(9) pertains for aspects of the o- and q-families (13.38)

Possibly, the 1/2 in item (13.34) correlates with 1 out of 2 of the following families - o and q. Above, we show that S=9/2 correlates with a limit on the q-family. [Discussion related to items including and following item (11.7)] 1 family - o - remains for symmetries larger than SU(9). The next item pertains.

| SU(11) pertains for aspects of the o-family (13.39)

Possibly, the 2/5 in item (13.35) correlates with 2 out of 5 of the following sets of oscillator indices - E11-E10-P10-P11, E9-E8-P8-P9, E7-E6-P6-P7, E5-E4-P4-P5, and E3-E2-P2-P3. In this paper, we do not further explore this matter. In any event, the next item pertains.

| The limit S≤5 pertains to the o-family (13.40)

We suggest research

SOR.13.1 Verify (to some confidence level) or refute that SU(7) symmetry correlates with observations.
STR.13.1 Better connect theoretically 2 concepts (channels and the non-applicability of a possible SU(9)-related symmetry) that point to a limit of S≤4 for the e-family.
STR.13.2 To what extent might people benefit by considering SU(5) to be a relevant symmetry larger than the Standard Model symmetry?

We list references

Ref.13.1 John Baez and John Huerta, The Algebra of Grand Unified Theories, *Bulletin of the American Mathematical Society, Volume 47,* Number 3, July 2010, pages 483-552. http://www.ams.org/journals/bull/2010-47-03/S0273-0979-10-01294-2 /

Section 14 Kinematics of e- and s-family bosons

Abs.14.1 IOM correlate with kinematics of e- and s-family bosons.

Context

We characterize, with respect to kinematics, some work above

Above, we do not much address details of quantum kinematics of basic particles or other objects.

We anticipate opportunities, with respect to kinematics, to extend work above

Here, we explore IOM that correlate with quantum kinematics. We correlate results with e- and s-family bosons.

Core

We explore aspects of an alternative QM-type-CS approach

Items including and following item (4.24) provide the basis for a QM-type-CS approach.

For D_P=3, the next items provide an alternative form of the Laplacian operator. Here, r_2 provides a radial coordinate in 2 dimensions, φ provides the related angular coordinate, and x provides a linear coordinate for the third dimension. We use coordinates y and z for a linear description correlating with r_2^2. [Ref.14.1]

$$\nabla^2 = r_2^{-1}(\partial/\partial r_2)(r_2)(\partial/\partial r_2) + r_2^{-2}(\partial^2/\partial^2\varphi) + (\partial^2/\partial^2 x) \quad (14.1)$$
$$r_2^2 = y^2 + z^2 \quad (14.2)$$

The next items restate item (14.1).

$$\nabla^2 = \nabla_2^2 + (\partial^2/\partial^2 x) \quad (14.3)$$
$$\nabla_2^2 = r_2^{-1}(\partial/\partial r_2)(r_2)(\partial/\partial r_2) + r_2^{-2}(\partial^2/\partial^2\varphi) \quad (14.4)$$

The next items provide relevant equations and solutions for $D_{P'}$=2.

$$\xi\,\Psi(r_2) = (\xi_0/2)\,(\,-\eta^2\,\nabla_2^2 + \eta^{-2}r_2^2\,)\,\Psi(r_2) \quad (14.5)$$
$$\nabla_2^2 = r_2^{-1}(\partial/\partial r_2)(r_2)(\partial/\partial r_2) - \Omega_2 r_2^{-2} \quad (14.6)$$

ξ and $\xi_0/2$ denote numbers (14.7)
$\Psi(r_2)$ denotes a wave function (14.8)
r_2 denotes a variable, with dimensions of length (14.9)
η denotes a length (14.10)
Ω_2 denotes a number (14.11)

$$V = (\xi_0/2)\,\eta^{-2}\,r_2^2 \quad (14.12)$$

Paralleling items (4.35) and (4.36), the next items characterize solutions. Here, we use $D_{P'}$=2. Item (14.17) follows from $\Omega_2 = \nu(\nu+D-2)$.

$$\xi = (2+2\nu)\,(\xi_0/2) \quad (14.13)$$
$$\Omega_2 = \nu(\nu+D_{P'}-2) = \nu(\nu+2-2) = \nu^2 \quad (14.14)$$
$$\nu = -1/2 \text{ or } -1 \quad (14.15)$$
$$\Omega_2 = \pm S'(S' + D_{P'} - 2) = \pm(S')^2 \quad (14.16)$$

- $D = 3 - \Omega_2$, for $\nu = -1$ (14.17)
- $D = 5/2 - 2\Omega_2$, for $\nu = -1/2$

For IOM(3,3), #E=3. One can parallel work above, based on substitutions such as those the next items show. Here, $\rightarrow$ denotes replaces.

$$\upsilon \rightarrow \xi \quad (14.18)$$
$$\upsilon_0 \rightarrow \xi_0 \quad (14.19)$$
$$t' \rightarrow r \quad (14.20)$$
$$\nabla_{t'}^2 \rightarrow \nabla_r^2 = \nabla^2 \quad (14.21)$$
$$\ddot{\iota} \rightarrow \eta \quad (14.22)$$
$$t \rightarrow x \quad (14.23)$$

The next item restates, in the above coordinates, item (4.24).

$$\upsilon\,\Psi(t') = (\upsilon_0/2)\,(\,-\ddot{\iota}^2\,\nabla_{t'}^2 + \ddot{\iota}^{-2}t'^2\,)\,\Psi(t') \quad (14.24)$$

The next item provides a QM-type-CS analog to item (4.17), which states 0 = Œ. Here, we use notation that recognizes that the relevant wave function is a function of both QE-like coordinates and QP-like coordinates. Here, we do not show angular coordinates.

$$0 = \upsilon\, \Psi(r,t') - \xi\, \Psi(r,t') \tag{14.25}$$

The next item defines a symbol with dimensions of velocity.

$$v = \eta \,/\, ï \tag{14.26}$$

The next items provide characteristics we posit for an edge solution. Here, δ denotes a Dirac delta function. Here, t and t' correlate with items (14.23) and (14.20), respectively.

$$\Psi \propto \delta(x-vt) \tag{14.27}$$
$$\Psi \propto \delta(r^2-v^2t'^2) \tag{14.28}$$
$$\upsilon_0 = \xi_0 \tag{14.29}$$

The next item pertains.

$$(\upsilon_0/2)\,(ï^{-2}t^2\,) - (\xi_0/2)\,(\eta^{-2}x^2\,) = 0 \tag{14.30}$$

The next items show relationships involving operators E and P corresponding to, respectively, energy and momentum for 1-dimensional motion.

$$E^2 \propto (\upsilon_0/2)\,(\,-ï^2(\partial^2/\partial t^2)\,) \tag{14.31}$$
$$v^2P^2 \propto (\xi_0/2)\,(\,-\eta^2(\partial^2/\partial x^2)\,) \tag{14.32}$$
$$\begin{aligned} E^2 - v^2P^2 &\propto (\upsilon_0/2)\,(\,-ï^2(\partial^2/\partial t^2)\,) - (\xi_0/2)\,(\,-\eta^2(\partial^2/\partial x^2)\,) \\ &\propto -\,(\upsilon_0/2)\,(\,\Omega_2(\text{QE-like})\,) + (\xi_0/2)\,(\,\Omega_2(\text{QP-like})\,) \\ &= (\xi_0/2)\,(\,\Omega_2(\text{QP-like}) - \Omega_2(\text{QE-like})\,) \end{aligned} \tag{14.33}$$

We make interpretations based on the next item.

$$v = c \tag{14.34}$$

For each of $D=D_{P'}=2$ and $D=D_{E'}=2$, the next items show the only edge solution. Here, $\nu = -1$.

S'	Ω_2	D	D+2ν	
				(14.35)
1	1	2	0	(14.36)

For these edge solutions, D+2ν=0, for each of QE-like and QP-like.

We derive $E^2-c^2P^2=0$ for the e-family

The next items show the number of Ω_2(QE-like) contributions and the number of Ω_2(QP-like) contributions that pertain for some e-family members. [Items following item (12.1), items following item (8.43), items following item (8.53), and items following item (12.9)]

Boson	Number of Ω_2(QE-like)	Number of Ω_2(QP-like)	(14.37)
2e2&	0	1	(14.38)
4e4&	0 or 1	2	(14.39)
6e6&	0 or 2	3	(14.40)
8e8&	0 or 3	4	(14.41)
4e24&	0	2	(14.42)
6e246&	0	3	(14.43)
8e2468&	0	4	(14.44)

Each Ω_2(QE-like) or Ω_2(QP-like) contribution contributes to K-mass an amount proportional to D+2ν=0. The next items pertain to each member of the e-family.

$$\text{K-mass} = 0 \quad (14.45)$$
$$E^2 - c^2P^2 = 0 \quad (14.46)$$

We derive $E^2-c^2P^2=0$ for the s-family

For s-family, the next items pertain for each of the 6 modes. [Items following item (10.1)] The Ω_2(QE-like) contribution contributes to K-mass an amount proportional to D+2ν=0. Items (14.45) and (14.46) pertain for the s-family.

Mode	Number of Ω_2(QE-like)	Number of Ω_2(QP-like)	(14.47)
2s(...)	1	0	(14.48)

Comments

We note some inside solutions

The next items show some inside solutions for ν=−1. Here, 2S' is an even integer. Here, D+2ν = 3−Ω_2 = $(S')^2$+1. [Item (14.17)]

S'	Ω_2	D	D+2ν	(14.49)
0	0	3	1	(14.50)
1	-1	4	2	(14.51)
2	-4	7	5	(14.52)
3	-9	12	10	(14.53)
4	-16	19	17	(14.54)
5	-25	28	26	(14.55)

The next items show some inside solutions for ν=−1/2. Here, 2S' is an odd integer. Here, D+2ν = 3/2 − 2Ω_2. [Item (14.17)]

S'	Ω_2	D	D+2ν	
1/2	1/4	2	1	(14.57)
1/2	−1/4	3	2	(14.58)
3/2	−9/4	7	6	(14.59)
5/2	−25/4	15	14	(14.60)
7/2	−49/4	27	26	(14.61)
9/2	−81/4	43	42	(14.62)

(14.56)

We list references

Ref.14.1 Wolfram Alpha, computational knowledge engine, Wolfram Alpha LLC, http://mathworld.wolfram.com/Laplacian.html.

Section 15 Kinematics of non-zero-mass basic particles

Abs.15.1 IOM correlate with kinematics of basic non-zero-mass particles.

Context

We characterize, with respect to kinematics, some work above

Above, we do not much address details of quantum kinematics of non-zero-mass basic particles.

We anticipate opportunities, with respect to kinematics, to extend work above

Here, we explore IOM that correlate with quantum kinematics for non-zero-mass basic particles.

Core

We discuss how to derive a kinematic equation related to the Dirac equation

Fields associated with basic bosons correlate with ν=−1 in IOM. Work above uses math associated with ν=−1 to derive results for $E^2-c^2P^2$ for zero-mass basic bosons. [Section 14] There, we show math that separates out 1 QE-like coordinate and 1 QP-like coordinate and then develops an appropriate solution. People might think of ν=−1 as being 1 dimension inside from a ν=−3/2 edge.

Fields for basic fermions correlate with ν=−1/2. People might think of ν=−1/2 as being 2 dimensions from a ν=−3/2 edge. For basic fermions, we consider parallels to work for basic bosons, but with 2 dimensions of QE-like partial differential equations correlating with motion and with 2 dimensions of QP-like partial differential equations correlating with motion. The next item pertains.

Gss.15.1 A merger of 2 sets (1 QE-like and 1 QP-like), each of 2 operators, correlates with a standard representation for $E^2-c^2P^2$ that people use based on the Dirac equation. Operators act on 4-component spinors. People can represent aspects of the operators via gamma matrices. (15.1)

We discuss matters related to mass for basic fermions

For #E=3=#P basic fermions, there remain 1 more dimension for QE-like oscillators and 1 more dimension for QP-like oscillators. For each of the QE-like case and the QP-like case, item (4.33) would pertain. We think that nothing quite as simple as the math leading to $E^2-c^2P^2=0$ for photons pertains. We think that the next items pertain.

Kinematic masses for basic fermions correlate with exponential functions (15.2)
- People might consider that remarks following item (4.49) correlate with this notion
- People might consider that items including and following item (11.22) correlate with this notion

Variables for the exponential-like functions may correlate with ... (15.3)
- Generation
  - People might consider the relevant variable here to be QP-like
  - For items including and following item (11.22), mass increases as generation increases
- Charge for M"=0 particles
  - People might consider the relevant variable here to be QE-like
  - For M"=0 elements in items including and following item (11.22), mass decreases as the magnitude of charge increases

Work in Section 18 considers relative sizes of masses for S=1/2 basic fermions.

We discuss how to derive a kinematic equation for w-family bosons

Work in Section 14 correlates, for the e-family, the 0 in $0=E^2-c^2P^2$ with an edge solution. The next item provides a basis for determining relative masses for w-family basic bosons.

Gss.15.2 W-family masses correlate with Ω_2-related $\nu=-1$ inside solutions. (15.4)

Work in Section 17 considers relative sizes of masses for w-, h-, and o-family basic bosons.

Comments

We discuss the sign of $E^2-c^2P^2$

Perhaps, for all basic particles, the next item pertains. Here, m denotes the kinetic mass people correlate with the particle.

$E^2 - c^2P^2 = \text{sign}(\Omega)\ |m^2|\ c^4$, in which (15.5)
- sign(>0) = 1
- sign(0) = ±1
- sign(<0) = −1

The next items show the mode correlating with sign(0) = −1. The mode cannot be excited (for the same reason that, for 2e2&, the P1 oscillator cannot be excited).

n_{EB} n_{EA} n_{E9} n_{E8} n_{E7} n_{E6} n_{E5} n_{E4} n_{E3} n_{E2} n_{E1} n_{P1} n_{P2} n_{P3} n_{P4} n_{P5}	Mode	(15.6)
−1 −1	0h1'	(15.7)

We know of nothing in observations or this paper that we think rules out item (15.5). We discuss implications below. [For example, discussion preceding item (17.29), and items following item (17.29); for example, item (23.1)]

We suggest research

STR.15.1 Show that IOM correlate with applicability, regarding basic fermions, of the Dirac equation.
STR.15.2 Determine the extent to which people can benefit by considering $E^2 - c^2P^2$ to be negative for the s- and q-families.
STR.15.3 Determine the extent to which people can benefit by considering $E^2 - c^2P^2$ to both positive and negative for the h-family.

Part 5 Particles and phenomena

Context

We discuss some unresolved physics

People say that people do not adequately understand mechanisms governing changes in the observed rate of expansion of the universe. People correlate near-flatness of the universe with the notion of a critical density of stuff. People say that models do not adequately correlate with the observed universe's having much more matter than antimatter. People say that the Standard Model does not adequately correlate with observed sizes of violations of symmetries related to CPT (charge, parity, and time) symmetry. People do not know masses for neutrinos. People say that models do not adequately correlate with neutrino oscillations. Perhaps, people would say that the Standard Model does not adequately correlate with masses of elementary particles.

We anticipate addressing such unresolved matters

We anticipate suggesting aspects of IOM that correlate with resolving such matters.

Core

We preview sections in this part

Section 16 correlates e-family coherences with changes in the rate of expansion of the universe and with lack of large-scale curvature of the universe.
Section 17 provides formulas linking masses of non-zero-mass basic bosons.
Section 18 provides approximate formulas linking masses of basic fermions.
Section 19 discusses mechanisms for various phenomena, including neutrino oscillations.
Section 20 estimates strengths for various e-family-mediated interactions.
Section 21 provides candidates for dark matter and candidates for dark-energy stuff.
Section 22 provides a mechanism leading to matter/antimatter imbalance. This section also discusses CPT-related symmetry violations and provides a basis for people's not being able to explain such violations fully via the Standard Model.
Section 23 discusses possibilities regarding $\Omega<0$ basic particles, uncertainty, and quasars.

Section 16 The universe's curvature and rate of expansion

Abs.16.1 E-family coherences provide for changes in the rate of expansion of the observed universe.
Abs.16.2 E-family coherences correlate with the universe's having zero large-scale curvature.

Context

We review some observations about the universe's rate of expansion and large-scale curvature

Traditional physics includes observations indicating that directly observed astrophysical objects move away from each other. People discuss a concept of a rate of expansion of the universe. People say that the rate varies during the history of the universe.

People deduce rates of expansion from observations of photons. People express approximate time after the big bang in terms of the redshift, z, people correlate with photons that people and equipment observe. Redshift z=0 pertains to photons emitted recently. Redshift numbers are positive. A value of z denotes a time before the time people associate with a smaller value of z.

The next items pertain. [Ref.16.1 and Ref.16.2] 3 eras exist.

For z > some number greater than 2.3, the observed rate of expansion increases (16.1)

- We call this era ZE1

During ~2.3 > z > ~0.7, the observed rate of expansion decreases (16.2)

- We call this era ZE2

During 0.46±0.13 > z > 0, the observed rate of expansion increases (16.3)

- We call this era ZE3

Traditional physics does not seem to include interactions that could cause such changes in the rate of expansion. Sometimes, people seem to correlate increasing rate of expansion with dark energy.

The next items correlate redshifts with times after the big bang. [Ref.16.3]

z	Time after the big bang (years)	
		(16.4)
2.3	2.9×10^9	(16.5)
0.7	7.3×10^9	(16.6)
0.46	8.9×10^9	(16.7)
0	13.8×10^9	(16.8)

Also, traditionally, people discuss the extent to which to consider that, on large scales, the universe exhibits positive curvature (or, is spherical, or $\Omega_0>1$), no curvature (or, is flat, or $\Omega_0=1$), or negative curvature (or, is hyperbolic, or $\Omega_0<1$). (Ω_0 is not related to Ω or to Ω_2.) Observations indicate that $\Omega_0\approx1$ may pertain. [Ref.16.4] People associate the notion of a critical density with $\Omega_0=1$.

We anticipate correlating 3 periods in the expansion with 3 e-family coherences and we anticipate a non-traditional explanation for $\Omega_0\approx1$

We anticipate correlating 8e2468&, 6e246&, and 4e24& with the 3 eras denoted by ZEï, with ï = 1, 2, and 3 respectively.

Also, we anticipate correlating these 3 coherences with the observed lack of large-scale curvature.

Core

We discuss curvature

Traditionally, people develop and use coordinate systems to discuss aspects of space time. The next item pertains.

Space time coordinate systems correlate with e-family mathematical physics (16.9)

People might not associate with each other concepts the next two items mention.

The term curvature	(16.10)
A fold people might associate with the Minkowski metric	(16.11)

We use the term flat-beyond-1-fold to apply to the Minkowski metric. Here, Minkowski metric can pertain to a space with at least 1 QE-like dimension and at least 1 QP-like dimension. We choose to use the term curvature for other purposes.

Traditional astrophysics includes the concept of co-moving coordinates. For objects situated adequately far apart, people might use combined-QE-like-and-QP-like geodesic-distance (not QP-like distance) to describe separation.

We discuss mechanisms affecting the rate of expansion

We think about 2 clumps of stuff. These clumps have similar size. The clumps neighbor each other.

The next items discuss forces that dominate within each clump and between the 2 clumps. These items reflect the observation that the universe expands. During such expansion, effects of an $R^{-2ύ}$ force (within or between the 2 clumps) evolve from being more than effects of $R^{-2(ύ-1)}$ forces to being less than effects of $R^{-2(ύ-1)}$ forces. The next items define 4 eras.

During era FE1, the force associated with 8e2468& dominates	(16.12)
During era FE2, the force associated with 6e246& dominates	(16.13)
During era FE3, the force associated with 4e24& dominates	(16.14)
During era FE4, the forces associated with 4e4& and 2e2& dominate	(16.15)

Today, observed atoms, planets, stars, galaxies, and galactic clusters exhibit era FE4 behavior. At and beyond a size that is larger than sizes people associate with galactic superclusters, people observe FE3 behavior.

The next items pertain.

Gss.16.1	For observed astrophysical objects of above some size, era FEó correlates with era ZEύ, for 1≤ó=ύ≤3.	(16.16)
Gss.16.2	8e2468& repels astrophysical objects from each other. 6e246& attracts astrophysical objects to each other. 4e24& repels astrophysical objects from each other.	(16.17)

We discuss the topic of curvature of the universe

Section 8 shows IOM(1,ι) representations for 2e2&, 4e24&, 6e246&, and 8e2468&. We call these coherences the maximal-% e-family bosons. There, ï = 3, 5, 7, and 9, respectively. For the ground state of each such #E=1 representation, n_{E1}=0. The next items pertain.

Gss.16.3	For observations expressed in terms of co-moving coordinates (with 1 QE-like dimension and 3 QP-like dimensions), people would say that each of the forces 2e2&, 4e24&, 6e246&, and 8e2468& does not contribute to any apparent curvature of space time.	(16.18)

For observations expressed in terms of co-moving coordinates (with 1 QE-like dimension and 3 QP-like dimensions), people would say that each of the forces 4e4&, 6e6&, and 8e8& contributes to curvature of space time (16.19)

- People might say that, generally, effects of 4e4& greatly outweigh effects of 6e6% and 8e8&

The next item pertains.

Gss.16.4 To the extent maximal-% e-family bosons dominated (throughout the past history of the universe) interactions between objects, people can consider those objects to be part of a universe for which $\Omega_0 \approx 1$. (16.20)

Possibly, only era-FE4 interactions based on 4e4& led to significant perceived curvature (based on observations that are based on #E=1 physics and co-moving coordinates). The next item pertains. People might say that the observed $\Omega_0 \approx 1$ does not depend on the density of the universe approximating a particular value.

IOM correlate with observed $\Omega_0 \approx 1$ (16.21)

Comments

We note possibilities that FE2 or FE1 behavior still dominate for some objects

Should the universe be adequately vast, large objects could still experience FE2 or FE1 behavior.

We suggest research

SOR.16.1 Estimate charges of objects for which 4e24& currently dominates.

We list references

Ref.16.1 N. G. Busca, et. al., Baryon Oscillations in the Lyα forest of BOSS quasars, arXiv:1211.2616 [astro-ph.CO].

Ref.16.2 A. Riess, et. al., Type Ia Supernova Discoveries at z > 1 from the *Hubble Space Telescope*: Evidence for Past Deceleration and Constraints on Dark Energy Evolution, *The Astrophysical Journal*, 607, 665 (2004). (doi:10.1086/383612) (http://iopscience.iop.org/0004-637X/607/2/665)

Ref.16.3 N. Gnedin, Cosmological Calculator for the Flat Universe. (http://home.fnal.gov/~gnedin/cc/)

Ref.16.4 NASA, http://map.gsfc.nasa.gov/universe/uni_shape.html

Section 17 W-, h-, and o-family masses and charges

Abs.17.1 IOM correlate with relative masses for w- and h-family bosons.
Abs.17.2 IOM may correlate with masses for o-family bosons.
Abs.17.3 Threshold energies for creating minimum numbers of o-family bosons may be (in units of mass) 241, 274, 526, ..., 1379 GeV/c^2.
Abs.17.4 The 2o3 boson has charge $+(1/3)|q_e|$ and the 2o2 boson has charge $-(1/3)|q_e|$.

Context

We do not know the extent to which traditional theory predicts masses for some non-zero-mass basic bosons

People state masses for w- and h-family particles. People measure lower bounds for masses of (hypothetical) leptoquarks. We do not know the extent to which traditional models predict leptoquark masses.

We anticipate developing approximate formulas relating masses of basic bosons

We explore the possibility that items following item (14.49) pertain to properties of w-, h-, and o-family bosons. We show a formula that approximates masses of w- and h-family basic bosons. We suggest an extrapolation of the formula that may correlate with o-family masses. We discuss possible charges for o-family basic bosons.

Core

We explore masses for w- and h-family basic bosons

The next items provide a formula that correlates with masses for w-family basic bosons (2w%) and the h-family basic boson (0h1). Here, m(ό) denotes the mass of basic boson ό.

$$m`` = m(\text{Z boson}) / 3 \quad (17.1)$$
$$(m(2w\%))^2 \approx (m``)^2 \times f(2w\%) \quad (17.2)$$
$$(m(0h1))^2 \approx (m``)^2 \times f(0h1) \quad (17.3)$$
$$f(2w1) = 9 \quad (17.4)$$
$$f(2w2) = f(2w3) \approx 7 \quad (17.5)$$
$$f(0h1) = 17 \quad (17.6)$$

The next items compare calculated and experimental masses. [Ref.17.1, Ref.17.2, and Ref.17.3]

Particle	Symbol	f($ï%)	Calculated mass (GeV/c^2)	Experimental mass (GeV/c^2)	(17.7)
Z	2w1	9	91.188	91.1876±0.0021	(17.8)
W	2w2, 2w3	7	80.420	80.385±0.015	(17.9)
Higgs	0h1	17	125.325	125.3 ± 0.4 (stat) ± 0.5 (sys) [Ref.17.2] 126.0 ± 0.4 (stat) ± 0.4 (sys) [Ref.17.3]	(17.10)

We correlate w- and h-family masses with IOM related to particle kinematics

Here (as we do for the e- and s-families [Section 14]), we equate values of D+2ν (related to Ω_2) with contributions to the squares of kinetic masses. The next items correlate with f($ï%)-column results that, respectively, items (17.8), (17.9), and (17.10) show.

The square of the mass of a Z boson correlates with the sum of (17.11)
- −1 = minus a $D_{E'}$=2 instance of item (14.50)
- +10 = plus a $D_{P'}$=2 instance of item (14.53)

The square of the mass of a W boson correlates with the sum of (17.12)
- −2 = minus a $D_{E'}$=2 instance of item (14.51)
- −1 = minus a $D_{E'}$=2 instance of item (14.50)
- +10 = plus a $D_{P'}$=2 instance of item (14.53)

The square of the mass of a Higgs boson correlates with the sum of (1 term) (17.13)
- +17 = plus a $D_{P'}$=2 instance of item (14.54)

Perhaps the item-(17.9) discrepancy regarding W-boson mass correlates with a possible non-zero magnetic moment for W bosons or with an aspect people might associate with an application of a perturbation theory.

We discuss charges for 2o% bosons

The next items note representations for ground states of 2o% bosons. [Items following item (12.25)] The value of each relevant $n_{P\acute{o}}$ is 0.

n_{E9} n_{E8} n_{E7} n_{E6} n_{E5} n_{E4} n_{E3} n_{E2} n_{E1} n_{P1} n_{P2} n_{P3} n_{P4} n_{P5} n_{P6} n_{P7} n_{P8} n_{P9}	Modes	(17.14)
0 # # 0 0 0	2o3	(17.15)
# 0 # 0 0 0	2o2	(17.16)
# # 0 0 0 0	2o1	(17.17)

Based on Œ=0, exciting an $n_{E\ddot{\imath}}$ from 0 to 1 requires exciting an $n_{P\acute{o}}$ from 0 to 1. We know of no preferred value of Pó.

The next item pertains to 2o2 and shows an amplitude for such an excitement for n_{E2}. Here, we use notation that denotes | n_{E3} , n_{E2} , n_{E1} , n_{P1} , n_{P2} , n_{P3} >.

$$(1/3)^{1/2} \mid \#, 1, \#, 1, 0, 0 > \quad (17.18)$$
$$+ (1/3)^{1/2} \mid \#, 1, \#, 0, 1, 0 >$$
$$+ (1/3)^{1/2} \mid \#, 1, \#, 0, 0, 1 >$$

The charge for an $n_{E\ddot{\imath}}$=0 item in the items following item (17.14) correlates with the charge associated with the Pï oscillator. The other 2 values of Pó are not relevant.

The next item pertains. Here, as above, Q' denotes charge divided by $|q_e|$.

$$Q'(2o2) \quad (17.19)$$
$$= ((1/3)^{1/2} < \#, 1, \#, 0, 1, 0 \mid) ((1/3)^{1/2} \mid \#, 1, \#, 0, 1, 0 >) Q'(2w2) + 0$$
$$= (1/3)\ Q'(2w2)$$

The next item pertains.

Q'(2oï) = (1/3) Q'(2wï), for ï = 3, 2, and 1 (17.20)

We discuss charges for o-family bosons $o% for which $ = 4, 6, 8, or 10

The next item pertains.

For $ = 4, 6, 8, or 10, Q'($oï) = 0, for ï = 1 and for ï≥3 (17.21)

The next items illustrate 2 possibilities for extending this work to predict charges for $oï for which $ = 4, 6, 8, or 10 and ï = 3. Item (17.23) exemplifies a possibility correlating with stating (in items following item (12.25)), $n_{P\acute{o}}$=# for each instance of Pó≥4 for which an item following item (12.25) states $n_{P\acute{o}}$=0. We call this possibility o-type-3. Item (17.24) exemplifies a possibility correlating with keeping as is items following item (12.25). We call this possibility o-type-11.

n_{E9}	n_{E8}	n_{E7}	n_{E6}	n_{E5}	n_{E4}	n_{E3}	n_{E2}	n_{E1}	n_{P1}	n_{P2}	n_{P3}	n_{P4}	n_{P5}	n_{P6}	n_{P7}	n_{P8}	n_{P9}	Modes	(17.22)
				#	#	0	#	#	0	0	0	#	#					4o3	(17.23)
				#	#	0	#	#	0	0	0	0	0					4o3	(17.24)

The next items summarize results for the o-type-3 possibility.

Q'($oï) = (1/3) Q'(2wï), for $ = 4, 6, 8, or 10 and ï = 3, 2, and 1 (17.25)
Values of Q' for o-family particles are ±1/3 and 0 (17.26)

The next items summarize results for the o-type-11 possibility.

Q'(4oï) = (1/5) Q'(2wï), for ï = 3, 2, and 1 (17.27)
Q'(6oï) = (1/7) Q'(2wï), for ï = 3, 2, and 1
Q'(8oï) = (1/9) Q'(2wï), for ï = 3, 2, and 1
Q'(10oï) = (1/11) Q'(2wï), for ï = 3, 2, and 1
Values of Q' for o-family particles are ±1/3, ±1/5, ±1/7, ±1/9, ±1/11, and 0 (17.28)

Based on work in Section 12 and Section 13 and on work below in this Section 17, perhaps o-type-11 pertains. In this paper, we do not make a choice between o-type-3 and o-type-11.

We discuss possible o-family masses and o-family creation threshold energies

The next items show how patterns pertaining to w- and h-family masses might pertain to o-family masses. Here, f parallels the f in item (17.7). The f column shows the sum of numbers in the columns to the left of the f column. We omit an E11, E10 column. That column would show blanks. The minimum number of bosons (min. numb. of bosons) column echoes concepts that o-family particles must be created in multiples [Item (9.72)]. We think the minimum numbers of bosons dovetail with o-type-11 (and do not violate the alternative possibility (o-type-3) regarding charges). The threshold column shows an estimated energy (stated in units GeV/c^2 correlating with mass) needed to create the particles. We compute the thresholds by assuming (for o-family particles) that zero contribution is needed regarding binding energy. (In contrast, for quarks, the binding energy is non-zero. For example, the mass of a proton exceeds the sum of the masses of the 3 quarks contained in the proton.) We explain the patterns (which we use for the leftmost 6 columns in these items, plus the omitted E11, E10 column) below. [Items including and following item (17.53)] The f=−17 possibility [Item (17.32)] correlates with 0h1'. [Item (15.7)]

E9, E8	E7, E6	E5, E4	E3, E2	E1, P1	P2, P3	f	Min. numb. of bosons	Threshold (GeV/c^2)	Bosons	(17.29)
				−1	10	9	1	91	2w1	(17.30)
			−2	−1	10	7	1	80	2w2, 2w3	(17.31)
				±17		±17	1	125	0h1	(17.32)
			−10	1		−9	3	274	2o1	(17.33)
			−10	1	2	−7	3	241	2o3, 2o2	(17.34)
		−5	−10	1		−14	5	569	4o1	(17.35)
		−5	−10	1	2	−12	5	526	4o3, 4o2	(17.36)
		−5	−10	1	2	−12	5	526	4o5, 4o4	(17.37)
	−2	−5	−10	1		−16	7	851	6o1	(17.38)
	−2	−5	−10	1	2	−14	7	796	6o3, 6o2	(17.39)
	−2	−5	−10	1	2	−14	7	796	6o5, 6o4	(17.40)
	−2	−5	−10	1	2	−14	7	796	6o7, 6o6	(17.41)
−1	−2	−5	−10	1		−17	9	1128	8o1	(17.42)
−1	−2	−5	−10	1	2	−15	9	1060	8o3, 8o2	(17.43)
−1	−2	−5	−10	1	2	−15	9	1060	8o5, 8o4	(17.44)
−1	−2	−5	−10	1	2	−15	9	1060	8o7, 8o6	(17.45)
−1	−2	−5	−10	1	2	−15	9	1060	8o9, 8o8	(17.46)
−1	−2	−5	−10	1		−17	11	1379	10o1	(17.47)
−1	−2	−5	−10	1	2	−15	11	1295	10o3, 10o2	(17.48)
−1	−2	−5	−10	1	2	−15	11	1295	10o5, 10o4	(17.49)
−1	−2	−5	−10	1	2	−15	11	1295	10o7, 10o6	(17.50)
−1	−2	−5	−10	1	2	−15	11	1295	10o9, 10o8	(17.51)
−1	−2	−5	−10	1	2	−15	11	1295	10o11, 10o10	(17.52)

The next items describe the patterns.

Integers (shown in the E9, E8 column through the P2, P3 column) reflect Ω_2-related values of D+2ν for (17.53)

- The S'=1 edge solution [Item following item (14.36)]
- The 0≤S'≤4 inside solutions [Items following item (14.49)]

The S'=1 edge solution pertains throughout for oscillator pairs E11, E10; P4, P5; P6, P7; P8, P9; and P10, P11 (17.54)

- This parallels work regarding oscillator pairs relevant to e-family masses [Items following item (14.37)]

The S'=1 edge solution pertains for elements for which items following item (17.29) show a blank (17.55)

For the o-family, ... (17.56)

- A QE-like instance of edge/S'=1 correlates with the E11-and-E10 pair
- A QE-like instance of inside/S'=0 correlates with the E9-and-E8 pair
- A QE-like instance of inside/S'=1 correlates with the E7-and-E6 pair
- A QE-like instance of inside/S'=2 correlates with the E5-and-E4 pair
- A QE-like instance of inside/S'=3 correlates with the E3-and-E2 pair
- A QP-like instance of inside/S'=0 correlates with the E1-and-P1 pair
- A QP-like instance of inside/S'=1 correlates with the P2-and-P3 pair
- A QP-like instance of edge/S'=1 correlates with the P4-and-P5 pair
- ...
- A QP-like instance of edge/S'=1 correlates with the P10-and-P11 pair

Items correlating with the w-family or the h-family reflect work above in Section 17 (17.57)

The next item restates a formula for estimating o-family kinetic masses.

K-mass = (m(Z boson) / 3) $|f|^{1/2}$ / (minimum number of bosons) (17.58)

The next items list estimated o-family kinetic masses.

Boson group	Mass (%≠1) (GeV/c^2)	Mass (%=1) (GeV/c^2)	(17.59)
2o%	80.4	91.2	(17.60)
4o%	105.3	113.7	(17.61)
6o%	113.7	121.6	(17.62)
8o%	117.7	125.3	(17.63)
10o%	117.7	125.3	(17.64)

Comments

We connect numbers of co-created non-zero-mass bosons with F for S≥3/2 fermions

We think that the next item pertains. [Items (11.49) and items following item (17.29)]

F = 1 + (minimum number of bosons) (17.65)

We suggest research

SOR.17.1 Determine properties (such as charge, mass, and magnetic moment) of o-family bosons.
SOR.17.2 Detect (or infer) or rule out (to some confidence level) the existence of o-family bosons with charges Q' (in units of $|q_e|$) of ±1/3 and 0,
SOR.17.3 Detect (or infer) or rule out (to some confidence level) the existence of o-family bosons with Q' of ±1/5, ±1/7, ±1/9 and ±1/11.
SOR.17.4 Measure masses for basic-boson o-family members and/or measure threshold energies for compound particles based on o-family members.
SOR.17.5 Determine or rule out (to some confidence level) non-zero binding energies for at-least-triplets of o-family bosons.
SOR.17.6 Verify or rule out (to some confidence level) that much of the difference between the W-boson mass we calculate and the observed W-boson mass correlates with a non-zero magnetic moment for W bosons.
STR.17.1 Estimate a value for the magnetic moment of W bosons.
STR.17.2 To what extent might people benefit by noticing that 2o% masses may equal counterpart 2w% masses and/or that 8o1, 10o1, and 0h1 masses may equal each other?

We list references

Ref.17.1 J. Beringer et al. (Particle Data Group), *PR D86*, 010001 (2012) and 2013 partial update for the 2014 edition (URL: http://pdg.lbl.gov). (http://pdg.lbl.gov/2013/tables/rpp2013-sum-gauge-higgs-bosons.pdf)
Ref.17.2 CMS collaboration (2012). "Observation of a new boson at a mass of 125 GeV with the CMS experiment at the LHC". *Physics Letters B* 716 (1): 30–61. arXiv:1207.7235. Bibcode:2012PhLB..716...30C. doi:10.1016/j.physletb.2012.08.021.
Ref.17.3 ATLAS collaboration (2012). "Observation of a New Particle in the Search for the Standard Model Higgs Boson with the ATLAS Detector at the LHC". *Physics Letters B* 716 (1): 1–29. arXiv:1207.7214. Bibcode:2012PhLB..716....1A. doi:10.1016/j.physletb.2012.08.020.

Section 18 S=1/2 basic fermion masses and q- and l-family charges

Abs.18.1 A formula approximates masses of quarks and charged leptons.
Abs.18.2 Neutrino masses or neutrino-mass-related math eigenvalues may be, in eV/c^2, approximately 1×10^{-1}, 2×10^{-4}, and 6×10^{-10}.

Context

We do not know the extent to which traditional theory predicts masses for basic fermions

Regarding the l-family, people discuss a formula [Item (18.100)] linking the masses of charged leptons. People do not know masses for neutrinos. People estimate upper bounds on neutrino masses.

Regarding the q-family, people measure approximate masses for quarks.

We anticipate developing approximate formulas correlating masses of basic S=1/2 fermions

We array l- and q-family masses in a way that suggests a pattern. We develop a formula that correlates with the pattern. Based on the possibility that the formula extends to neutrinos, on symmetry we correlate with neutrinos, and on upper bounds for neutrino masses, we suggest either possible masses for neutrinos or possible neutrino-mass-related math eigenvalues.

Core

We discuss q- and l- family charges

For S=1/2 basic particles other than neutrinos, items following item (11.22) list the charges Q'.

For S=1/2 neutrinos, Q'=0.

For S>1/2 basic particles we denote by 3q(...) or by 7q(...) or by 9q(...), $n_{E3}=n_{E2}=n_{P2}=n_{P3}=-1$ in the interaction representation. These particles do not interact directly with W bosons (2w2 and 2w3). These particles do not interact directly with charged o-family bosons ($o3 and $o2, for $ = 2, 4, 6, 8, or 10). These particles have no charge.

For S>1/2 basic particles we denote by combinations of $qa(...) and $qb(...), the charge Q'=0 pertains (by analogy with results for neutrinos).

We show formulas that correlate with S=1/2 q-family masses and with charged l-family masses

Edge solutions correlate with q- and l-family particles. Above, we discuss possibilities that considering ψ(r), with r<0, makes sense regarding such particles. [Item (4.49)] We note possibilities for roles for exponential and trigonometric functions.

The next item defines notation regarding items following item (11.22).

| m(M",M') denotes a calculated mass correlating with a position in a table that includes positions for S=1/2 charged basic fermions (18.1)

The next item pertains to trends in items following item (11.22).

| Gss.18.1 For charged leptons (either M'=−3 or M'=+3), people can benefit by correlating the range −1≤M"≤3 with an L=2 system. (18.2)

The next item pertains. For the L=2 system, −2≤M"−1≤+2. (We think that no solution with d(0)=−d(2) appeals. We think such a solution would correspond to a trigonometric function that is anti-symmetric with respect to M"−1. The next item features an expression that corresponds to a symmetric trigonometric function.)

| Gss.18.2 For the L=2 system that includes charged leptons, m(M",−3) ∝ $e^{M''\zeta''}(1+d(M''))$, in which −1≤M"≤3, d(0)=d(2), and d(−1)=d(1)=d(3)=0. (18.3)

The next items show results for M'=−3 charged leptons and related values of M". Here, we use the experimental masses for the electron and muon to calculate ζ". Then, we use an experimentally acceptable calculated mass for the tauon [Item (6.47)] to calculate m`. Then, we calculate d(0).

$$m(M'',-3) = m` \times \exp((M''+1)\zeta'') \times (1+d(M'')) \quad (18.4)$$
$$\zeta'' = (1/2)\log(m_{muon}/m_e) \approx 2.665799 \quad (18.5)$$
$$m` = m_{tauon} / \exp(4\zeta'') \approx 4.155987\times10^{-2}\ \mathrm{MeV}/c^2 \quad (18.6)$$
$$1 + d(0) = m_e / (m` \exp(\zeta'')) \quad (18.7)$$
$$d(2) = d(0) \approx -0.144926 \quad (18.8)$$

$$d(-1) = d(1) = d(3) = 0 \qquad (18.9)$$

The next item provides a calculated number not directly correlated with a particle.

$$m(1,-3) \approx 8.59326 \text{ MeV}/c^2 \qquad (18.10)$$

The next items correlate with a possible role for trigonometric functions.

$$d(M'') \approx d'' \times (1/2) \times (\cos(M''\pi) + 1) \qquad (18.11)$$

$$d'' = d(2) \approx -0.144926 \qquad (18.12)$$

We guess that the next item provides a useful set of m(M",0). Here, α denotes the fine-structure constant. No known particles correlate with M"≥0 and M'=0.

$$m(M'', 0) \approx m(M'', -3) \times \exp(\ (1/4) \log(1/\alpha)\ 3(1+M'')\ ) \qquad (18.13)$$

The next items show numbers.

$$m(0, 0) \approx 2.05\times10^{1} \text{ MeV}/c^2 \qquad (18.14)$$

$$m(1, 0) \approx 1.38\times10^{4} \text{ MeV}/c^2 \qquad (18.15)$$

$$m(2, 0) \approx 6.79\times10^{6} \text{ MeV}/c^2 \qquad (18.16)$$

$$m(3, 0) \approx 4.57\times10^{9} \text{ MeV}/c^2 \qquad (18.17)$$

The next items finish defining the formula correlating with approximate masses for quarks and charged leptons. We specify d(M",M') adequate to fit known experimental results. (For each M", we could use a trigonometric function to specify d(M", M').)

- $m(M'', M') \approx m(M'', 0) \times \exp(\ (1/4) \log(\alpha)\ (1+M'')\ |M'|\ ) \times (1+d(M'',M'))$ (18.18)
- $d(M'', \pm2) = -d(M'', \pm1) \approx d`(-d``)^{M''}$, for $0 \le M'' \le 2$ and $2 \ge |M'| \ge 1$ (18.19)
- $d(M'', M') = 0$ otherwise
- $d` \approx 0.2$ (18.20)
- $d`` \approx 0.4$ (18.21)

The next items pertain.

$$-d(0, -1) = d(0, -2) \sim 0.2 \qquad (18.22)$$

$$-d(1, -1) = d(1, -2) \sim -0.08 \qquad (18.23)$$

$$-d(2, -1) = d(2, -2) \sim 0.032 \qquad (18.24)$$

$$d(M'', M') = 0, \text{ otherwise within } 0 \le M'' \le 3,\ 3 \ge |M'| \ge 0 \qquad (18.25)$$

The next 2 sets of items compare calculated numbers with experimental numbers. [Ref.6.3 and Ref.18.1] The items show masses in units of MeV/c^2. For each particle, the bottom number (calc) comes from our calculations. Except regarding M"=2 quarks, the upper number (exp) comes from experiments. For M"=2 quarks, Ref.18.1 provides two possible ranges for quark masses. The upper range is based on mass-independent subtraction scheme(s) (MS). For M"=2, M'=−2, the first mass is the MS running mass and the second mass is the 1S mass. For M"=2, M'=−1, the first mass is labeled MS from cross-section measurements and the second mass is from direct measurements.

		M'	−3	...	...	
M"						(18.26) (18.27)
0	exp		0.510998928±0.000000011			(18.28)
0	calc		0.510998928 MeV/c^2			(18.29)
1	exp					(18.30)
1	calc					(18.31)
2	exp		105.6583715±0.0000035			(18.32)
2	calc		105.6583715			(18.33)
3	exp		1776.82±0.16			(18.34)
3	calc		1776.81			(18.35)

		M'	...	−2	−1	
M"						(18.36) (18.37)
0	exp			$2.3^{+0.7}_{-0.5}$	$4.8^{+0.7}_{-0.3}$	(18.38)
0	calc			2.10 MeV/c^2	4.79	(18.39)
1	exp			95±5	$(1.275\pm0.025)\times10^3$	(18.40)
1	calc			92.5	1.272×10^3	(18.41)
2	exp			$(4.18\pm0.03)\times10^3$	$(160^{+5}_{-4})\times10^3$	(18.42)
				$(4.65\pm0.03)\times10^3$	$(173.5\pm0.6\pm0.8)\times10^3$	
2	calc			4.367×10^3	164.1×10^3	(18.43)
3	exp					(18.44)
3	calc					(18.45)

Possibly, m(M",M') correlates with each of the 9 masses for quarks and charged leptons. People might say that we use the following numbers to fit the data.

$$m`, \zeta", d(0), \alpha, d`, \text{and } d`` \quad (18.46)$$

The next item correlates with work above. Here, we assume d(−1,±3)=d(−1,0)=0.

$$m(-1,-3) = m(-1,0) = m(-1,+3) = m` \quad (18.47)$$

We note symmetries that differentiate neutrinos from charged leptons

Items following items (11.33) and items following item (11.40) show representations that seemingly equate positron and neutrino-a and seemingly equate electron and neutrino-b.

The next items show alternative representations for generation-1 neutrinos. [Items following item (11.40)] Here, we show a symmetry that (unlike work in Section 11) does not correlate with QE-like aspects of interactions of which we know. (We discuss possible implications (of this symmetry) regarding interactions in Section 23.) These items show a QE-like SU(3) symmetry similar to that item (12.11) shows.

n_{E9}	n_{E8}	n_{E7}	n_{E6}	n_{E5}	n_{E4}	n_{E3}	n_{E2}	n_{E1}	n_{P1}	n_{P2}	n_{P3}	n_{P4}	n_{P5}	n_{P6}	n_{P7}	n_{P8}	n_{P9}	Concept	(18.48)
						−2	−2	−2	−3	−2	−1							neutrino-a	(18.49)
						−2	−2	−2	−3	1	2							neutrino-b	(18.50)

The next items indicate possible alternative representations of generation-1 charged leptons. [Items following item (11.33)] Here, we show a symmetry that (unlike work in Section 11) does not correlate with QE-like aspects of interactions of which we know. (We discuss possible implications (of this symmetry) regarding interactions in Section 23.) These items show a QE-like SU(2)×U(1) symmetry similar to that items following item (12.16) exhibit.

n_{E9}	n_{E8}	n_{E7}	n_{E6}	n_{E5}	n_{E4}	n_{E3}	n_{E2}	n_{E1}	n_{P1}	n_{P2}	n_{P3}	n_{P4}	n_{P5}	n_{P6}	n_{P7}	n_{P8}	n_{P9}	Concept	(18.51)
						−1	−2	−3	−3	−2	−1							positron	(18.52)
						−2	−1	−3	−3	−1	−2							electron	(18.53)

The next items pertain. Here, we add quarks by analogy to charged leptons. People consider that U(1) has 2 generators. People consider that SU(2) has 3 generators. Here, 6=2×3. People consider that SU(3) has 8 generators.

S=1/2 basic fermions	QE-like symmetry	Number of generators associated with the symmetry	(18.54)
neutrinos	SU(3)	8	(18.55)
charged leptons	SU(2)×U(1)	6	(18.56)
quarks	SU(2)×U(1)	6	(18.57)

We discuss a possible interpretation of the presence of β^6 in a ratio of vertex strengths

Work above correlates with a range of lepton masses that runs from m_e to βm_e. Work above correlates with a ratio of vertex strengths running from β^{-6} to β^0. [For example, item (6.2)] Work above points to a possible SU(2)×U(1) symmetry with 6 generators.

The next items may have relevance. Here, we show 6 groups of particles. Each group corresponds to 1 of the 6 instances that correlates with an SU(2)×U(1) symmetry. We label the 6 groups via Móï, as well as via the ("DMF") integers 0, 1, 2, ... 5. The concept columns extend cyclic labelling, based on names of leptons for which 0≤M"≤3. Here, positron denotes a standard term for a-electron. Here, a- denotes anti.

M"	concept	ó=1 for U(ó)	ó=2 for U(ó)	concept	ï for SU(2):ï	group ("DMF")	(18.58)
0	electron	M11			1	0	(18.59)
1					1		(18.60)
2	muon	M11			1	0	(18.61)
3	tauon	M11	M21	a-tauon	1	0, 1	(18.62)
4					1		(18.63)
5			M21	positron	1	1	(18.64)
6	muon	M12	M21	a-muon	1, 2	1, 2	(18.65)
7					2		(18.66)
8	tauon	M12			2	2	(18.67)
9	electron	M12	M22	positron	2	2, 3	(18.68)
10					2		(18.69)
11			M22	a-muon	2	3	(18.70)
12	tauon	M13	M22	a-tauon	2, 3	3, 4	(18.71)
13					3		(18.72)
14	electron	M13			3	4	(18.73)
15	muon	M13	M23	a-muon	3	4, 5	(18.74)
16					3		(18.75)
17			M23	a-tauon	3	5	(18.76)
18			M23	positron	3	5	(18.77)

Per channel, a gravitational (4e4&) interaction between 2 M"=18 charged leptons would be identical in magnitude to an electromagnetic (2e2&) interaction between the same particles. [For example, item (6.2)]

We extend work to include neutrinos

Work above points to a possible SU(3) symmetry with 8 generators. Work above correlates neutrinos with such a symmetry. [Item (18.55)]

Paralleling work just above, the set M" = {−6, −5, −4, −3} provides candidate values of M" for neutrinos.

To the extent the pattern in the concept column continues from items following item (18.58), the leftmost concept column for the next items pertains. For example, M"=−6 would correspond to a tauon-neutrino. That pattern possibly correlates with an SU(2)×U(1) symmetry. Here (for the next items), the rightmost 4 columns relate to the possible SU(3) symmetry (and, below, to items such as items (21.16) and (21.33)). Neutrino masses could follow the low-to-high order of charged-lepton masses, as the second concept column in the next items shows. Or, the (calculated) masses could represent mathematical eigenvalues that do not necessarily match with specific ï-neutrinos (for ï = electron, muon, and tauon).

M"	concept	ï for SU(3):ï	group ("DEF")	concept	concept	(18.78)
−6	tauon	1	0	electron	not a specific neutrino	(18.79)
−5		1				(18.80)
−4	electron	1	0	muon	not a specific neutrino	(18.81)
−3	muon	1, 2	0, 1	tauon	not a specific neutrino	(18.82)
...						(18.83)
15	muon	7, 8	6, 7			(18.84)
16		8				(18.85)
17	tauon	8	7			(18.86)
18	electron	8	7			(18.87)

Per channel, a gravitational (4e4&) interaction between 2 M"=18 neutrinos would be identical in magnitude to a hypothetical electromagnetic (2e2&) interaction between the same particles. There is no such electromagnetic interaction. (Here, we ignore the possibilities for electromagnetic interactions based on, for example, virtual pairs of W mesons.)

We discuss possible masses for neutrinos

The next item pertains.

Gss.18.3 The formula for m(M", M') has meaning for M"<−1. The trigonometric-like pattern for d(M") continues throughout the range −6≤M"≤3. d(M",0) = 0. (18.88)

The next items provide estimated values related to neutrinos. The items show masses in units of eV/c^2.

	M'	0	Possible correlation	
M"				(18.89)
				(18.90)
−6	exp			(18.91)
−6	calc	6×10^{-10} eV/c^2	mass-related math eigenvalue, electron-neutrino, or tauon-neutrino	(18.92)
−5	exp			(18.93)
−5	calc	4×10^{-7}	-	(18.94)
−4	exp			(18.95)
−4	calc	2×10^{-4}	mass-related math eigenvalue, muon-neutrino, or electron-neutrino	(18.96)
−3	exp			(18.97)
−3	calc	1×10^{-1}	mass-related math eigenvalue, tauon-neutrino, or muon-neutrino	(18.98)

Comments

We note that calculated neutrino masses seem consistent with observations

People interpret observations as implying that no more than 3 neutrinos (plus 3 possible anti-neutrinos) exist. [Ref.18.2] Work above regarding generations correlates with exactly 3 neutrinos (plus, if applicable, 3 antiparticles). People interpret observations as implying that the sum of the neutrino masses does not exceed 0.28 eV/c^2. [Ref.18.3 and Ref.18.4] Work above in this section is consistent with this limit. We do not know of relevant experimental results for neutrino masses.

We point to an involvement of square roots of masses

For charged leptons, the next item pertains. [Items (18.4) and (18.5)] The formula involves integer powers of the square roots of 2 masses.

$$m(M'', M') \approx m` \times ((m_{muon} / m_e)^{(1/2)})^{(M''+1)} \times (1+d(M'')) \quad (18.99)$$

The next item shows the Koide formula. People correlate this formula with aspects of the Standard Model.

$$(m_e + m_{muon} + m_{tauon}) / (m_e^{1/2} + m_{muon}^{1/2} + m_{tauon}^{1/2})^2 \approx 2/3 \quad (18.100)$$

The next item reflects nominal numbers we use, including the calculated value for m_{tauon}. The uncertainty-range may not be accurate.

$$(m_e + m_{muon} + m_{tauon}) / (m_e^{1/2} + m_{muon}^{1/2} + m_{tauon}^{1/2})^2 \approx 0.66666(\sim3) \quad (18.101)$$

We discuss possible correlations between powers of β and interaction strengths

The next items correlate with items (6.3) and (6.6). In particular, in item (18.104), γ=5 for tauons.

Force	Channels	Electron relative vertex strength per channel	Electron γ for $\beta^{-\gamma}$	Tauon relative vertex strength per channel	Tauon γ for $\beta^{-\gamma}$	(18.102)
2e2&	4	$1 = \beta^{-0}$	0	$1 = \beta^{-0}$	0	(18.103)
4e4&	3	β^{-6}	6	β^{-5}	5	(18.104)

The next items include information about the muon.

Particle	Symbol for mass m(M",M')	mass/m_e	M"	(18.105)
electron	$m(0,-3) = m_e$	$\beta^{0/3}$	0	(18.106)
-			1	(18.107)
muon	m(2,−3)	$\sim\beta^{(2/3)\cdot(1-0.02)}$ or $\sim 0.9\cdot\beta^{2/3}$	2	(18.108)
tauon	m(3,−3)	$\beta^{3/3}$	3	(18.109)

We note a possible correlation with inside solutions

The next item pertains. [Item (18.18)]

$$m(M'', M') \sim m(M'', 0) \times ((\alpha)^{-1/4(1+M'')})^{-|M'|}, \text{ for } 3 \geq |M'| \geq 1 \qquad (18.110)$$

The next items possibly correlate values of |M'| with Ω_2-related values of D+2ν. [Items including and following item (14.56)]

|M'|		(18.111)
1	D+2ν for Ω_2=1/4	(18.112)
2	D+2ν for Ω_2=−1/4	(18.113)
3	(D+2ν for Ω_2=1/4) + (D+2ν for Ω_2=−1/4)	(18.114)

We suggest research

SOR.18.1 Measure neutrino masses.

STR.18.1 To what extent does the appearance, in a formula for the ratios of masses of charged leptons, of a power of the ratio of the square roots of 2 lepton masses correlate with the possible applicability of the Koide formula? [Items (18.99) and (18.100)]

STR.18.2 Determine the extents to which numbers we state as correlating with neutrinos represent mass-related math eigenvalues and/or neutrino masses. [Items following item (18.89)]

We list references

Ref.18.1 J. Beringer et al. (Particle Data Group), *Phys. Rev. D86*, 010001 (2012). (http://pdg.lbl.gov/2012/tables/rpp2012-sum-quarks.pdf)

Ref.18.2 J. Beringer et al. (Particle Data Group), *PR D86*, 010001 (2012) and 2013 partial update for the 2014 edition (URL: http://pdg.lbl.gov). (http://pdg.lbl.gov/2013/tables/rpp2013-sum-leptons.pdf)

Ref.18.3 S. Thomas, F. Abdalla, and O. Lahav, Upper Bound of 0.28 eV on the Neutrino Masses from the Largest Photometric Redshift Survey, *Phys. Rev. Lett. 105*, 031301, 2010. (http://arxiv.org/abs/0911.5291)

Ref.18.4 A. Melchiorri, Constraints on Neutrino Physics from Planck, European Space Agency, http://www.rssd.esa.int/SA/PLANCK/docs/eslab47/Session06_CMB_Cosmology_and_Fundamental_Physics/47ESLAB_April_04_17_30_Melchiorri.pdf.

Section 19 Examples of interactions

Abs.19.1 We illustrate interactions involved in fermion-anti-fermion annihilation.
Abs.19.2 We illustrate interactions involved in neutrino oscillations.
Abs.19.3 We illustrate mechanics of channels.

Context

We discuss in this paper various interactions

Above, we allude to various types of interactions.

We illustrate various types of interactions

Here, we show examples of interaction vertices and types of interactions.

Core

We discuss a weak-interaction vertex

The next items pertain to a vertex in which an electron-neutrino and a W^- produce an electron. The description does not depend on choosing an n-type for the neutrino.

- An electron-neutrino enters (19.1)
- A W^- enters

$$| n_{E1}=2 , n_{P1}=\# , n_{P2}=1 , n_{P3}=\# >$$

- An electron exits (19.2)
- A w-family ground state remains

$$| n_{E1}=1 , n_{P1}=0 , n_{P2}=0 , n_{P3}=0 >$$

We discuss electron-positron annihilation

The dominant mode for electron-positron annihilation produces 2 photons. In a traditional Feynman diagram, an electron enters, 2 photons exit, and the positron exits as a continuation of the electron. People might say that the positron goes backward in time.

In an IOM treatment, the next items pertain.

- An electron enters (19.3)
- A positron enters

- The field associated with charged leptons absorbs each of the 2 charged leptons (19.4)
- 2 entangled photons are produced (19.5)
- The 2 entangled photons exit

The next items describe a possible scenario underlying the annihilation reaction.

Four virtual W bosons are created in the form (19.6)

$| n_{E1}=3 , n_{P1}=\# , n_{P2}=2 , n_{P3}=\# >$ plus $| n_{E1}=3 , n_{P1}=\# , n_{P2}=\# , n_{P3}=2 >$

Each of the electron and positron absorbs a W boson, resulting in the following (19.7)

- The field associated with charged leptons absorbs the 2 charged leptons
- The w-family state becomes

  $| n_{E1}=2 , n_{P1}=\# , n_{P2}=1 , n_{P3}=\# >$ plus $| n_{E1}=2 , n_{P1}=\# , n_{P2}=\# , n_{P3}=1 >$

The remaining 2 W bosons behave as (19.8)

$| n_{E1}=3 , n_{P1}=0 , n_{P2}=1 , n_{P3}=1 >$

The $| n_{E1}=3 , n_{P1}=\# , n_{P2}=1 , n_{P3}=1 >$ transforms into (19.9)

$| n_{E1}=2 , n_{P1}=-1 , n_{P2}=1 , n_{P3}=1 >$

The $| n_{E1}=2 , n_{P1}=-1 , n_{P2}=1 , n_{P3}=1 >$ consists of a pair of entangled photons (19.10)

Suppose one observer measures 1 of the photons to be $| n_{E1}=1 , n_{P1}=-1 , n_{P2}=1 , n_{P3}=0 >$ (or, P2-polarized). Then, an observer measuring the other photon measures $| n_{E1}=1 , n_{P1}=-1 , n_{P2}=0 , n_{P3}=1 >$ (or, P3-polarized).

People might say that there is no need to describe a particle as moving backward in time.

The field associated with charged leptons can participate in electron-positron pair production. For example, the reverse of above reaction can occur.

We discuss neutrino oscillations

People report that neutrinos oscillate. For example, a mu-neutrino can become a tau-neutrino. (Here, we simplify discussion by not discussing states that have non-zero amplitudes for each of 2 or 3 types of neutrinos.) People say that the presence of mass triggers oscillations or enhances rates of oscillation. [Ref.19.1]

The next items describe a neutrino-oscillation event. (Possibly, the reaction requires additional reactants in order to conserve energy and momentum.)

- A mu-neutrino enters (19.11)
- A 4e4 state exists (here, $\upsilon>0$ provides an occupation number)

  $| n_{E1}=1+\upsilon , n_{P1}=-1 , n_{P2}=\# , n_{P3}=\# , n_{P4}=\upsilon , n_{P5}=\# >$
- A 2o1 state exists (here, $n\geq 0$ provides an occupation number)

  $| n_{E3}=\# , n_{E2}=\# , n_{E1}=n , n_{P1}=1+n >$
- A unit of 4e4 converts to a unit of 2o1 (19.12)

  $| n_{E1}=1+\upsilon , n_{P1}=-1 , n_{P2}=\# , n_{P3}=\# , n_{P4}=\upsilon , n_{P5}=\# >$

  plus $| n_{E3}=\# , n_{E2}=\# , n_{E1}=n , n_{P1}=1+n >$

  $\rightarrow$

  $| n_{E1}=\upsilon , n_{P1}=-1 , n_{P2}=\# , n_{P3}=\# , n_{P4}=\upsilon-1 , n_{P5}=\# >$

  plus $| n_{E3}=\# , n_{E2}=\# , n_{E1}=n+1 , n_{P1}=2+n >$

- The neutrino increases its generation via an interaction with the unit of 2o1 (19.13)
- The 2o1 depopulates by 1 and becomes

  $| n_{E3}=\#, n_{E2}=\#, n_{E1}=n, n_{P1}=1+n >$
- A tau-neutrino exits (19.14)
- A 4e4 state exists (here, $\upsilon-1\geq0$ provides an occupation number)

  $| n_{E1}=\upsilon, n_{P1}=-1, n_{P2}=\#, n_{P3}=\#, n_{P4}=\upsilon-1, n_{P5}=\# >$
- A 2o1 state exists (here, $n\geq0$ provides an occupation number)

  $| n_{E3}=\#, n_{E2}=\#, n_{E1}=n, n_{P1}=1+n >$

The value of υ correlates with the existence somewhere of PROPE4. Based on concepts pertaining to lasing, the larger is υ, the higher is the probably (per unit time) of such an oscillation. The overall reaction does not change n for the 2o1 state. People might say that the 2o1 participates virtually.

Other similar reactions provide for oscillations between other pairs of types of neutrinos.

We illustrate mechanics of channels

The next item restates an aspect regarding channels. [Section 8]

channels(2e2&) : channels(4e4&) : channels(6e6&) : channels(8e8&) (19.15)

=

4 : 3 : 2 : 1

Each of the next items lists a possible set of channels relevant to 2e2&.

- P4-and-P5, P6-and-P7, P8-and-P9, P10-and-P11 (19.16)
- E9-and-E8, E7-and-E6, E5-and-E4, E3-and-E2, plus the channels item (19.16) lists (19.17)

The next item defines the term r/chan

r/chan(ï) = 4, 3, 2, and 1, respectively for ï = 2e2&, 4e4&, 6e6&, and 8e8& (19.18)

The next items pertain to creation, from a ground state, of an even-polarized $e%& e-family boson. Here, the list % has $-n_{P1}$ elements. We illustrate cases for which a Pυ-and-P(υ+1) channel participates.

- Aspect 1 (19.19)
  - For each of the $-n_{P1}$ even-numbered $ï\geq2$ for which $n_{Pï}=0$ (and $n_{Pï}\neq\#$), $n_{Pï}$ increases by 1
  - For 1 closed oscillator pair (say $n_{P\upsilon}$-and-$n_{P(\upsilon+1)}$, with υ being an even number),
    - The oscillator pair opens as $| n_{P\upsilon}=0, n_{P(\upsilon+1)}=-1 >$
    - $n_{P\upsilon}$ changes from 0 to n_{P1} ($n_{P\upsilon}$ decreases)
- Aspect 2 (19.20)
  - n_{E1} increases by $-n_{P1}$ (n_{E1} increases)
  - $n_{P\upsilon}$ changes from n_{P1} to 0 ($n_{P\upsilon}$ increases)
  - The $n_{P\upsilon}$-and-$n_{P(\upsilon+1)}$ oscillator pair closes

The next items pertain.

In aspect 1, $n_{P\upsilon}$ transits from non-negative to negative (19.21)
In aspect 2, $n_{P\upsilon}$ transits from negative to non-negative (19.22)

Comments

We discuss concepts related to channels

The next items provide candidate concepts for channels. Each features a pair of 2 adjacent oscillators. Each features Œ=0.

For ó an even positive integer, (19.23)
- $(1/2)^{1/2}$ (| $n_{E(ó+1)}$=−1 , $n_{Eó}$=0 > + | $n_{E(ó+1)}$=0 , $n_{Eó}$=−1 >)

For ó an even positive integer, (19.24)
- $(1/2)^{1/2}$ (| $n_{Pó}$=−1 , $n_{P(ó+1)}$=0 > + | $n_{Pó}$=0 , $n_{P(ó+1)}$=−1 >)

We do not further discuss choices of representations for e-family channels. We do not discuss the possible applicability of channels for basic bosons in the s-, w-, h-, and o-families.

We suggest research

SOR.19.1 Determine dependences of neutrino-oscillation rates on influences of mass or gravity.
STR.19.1 Determine the extent to which people might benefit by further exploring numbers and mechanics related to channels.

We list references

Ref.19.1 J. Beringer et al. (Particle Data Group), *Phys. Rev. D86*, 010001 (2012). "13. Neutrino mass, mixing, and oscillations," page 46. (http://pdg.lbl.gov/2012/reviews/rpp2012-rev-neutrino-mixing.pdf)

Section 20 E-family interaction strengths

Abs.20.1 Formulas provide approximate relative strengths for interactions mediated by e-family basic bosons and some e-family coherences.

Context

We note the extent to which we describe e-family interactions above

Above, we describe relative interaction strengths for the first 2 members of the photon-graviton series of 4 basic bosons. The work discusses interactions between 2 electrons.

We anticipate extrapolating to other relative interaction strengths

Here, we indicate approximate interaction strengths for interactions carried by e-family members other than just 2e2& and 4e4&.

Core

We discuss strengths of photon-graviton series basic bosons

The next item pertains. The M"=0 leptons are electrons and positrons.

Gss.20.1 For photon-graviton series basic bosons, for interactions between 2 M"=0 leptons, the relative vertex strength per r/chan follows a pattern established by the relative vertex strengths per r/chan for photons and gravitons. (20.1)

The next items show the pattern. For $e%& with 4≤$≤8, each relative vertex strength per r/chan equals β^{-6} times the preceding one.

Interaction	R/chans	Relative vertex strength per r/chan	Relative interaction strength per r/chan	(20.2)
2e2&	4	$1 = \beta^{-0}$	$1 = (\beta^{-0})^2$	(20.3)
4e4&	3	β^{-6}	$(\beta^{-6})^2$	(20.4)
6e6&	2	β^{-12}	$(\beta^{-12})^2$	(20.5)
8e8&	1	β^{-18}	$(\beta^{-18})^2$	(20.6)

Comments

We discuss magnitudes of strengths of some e-family interactions based on basic bosons

The next items provide numbers for relative vertex strengths and relative interaction strengths for interactions between 2 electrons, 2 positrons, or 1 electron and 1 positron. The second column notes logarithms of strengths relative to the strength per r/chan of a vertex for electromagnetism. The third column notes logarithms of strengths relative to the strength per r/chan of a vertex for gravity. The fifth column notes logarithms of interaction strengths relative to the strength of electromagnetism. The rightmost column notes logarithms of interaction strengths relative to the strength of gravity.

Interaction	Log_{10} (relative vertex strength per r/chan)	Log_{10} (relative vertex strength per r/chan)	R/chans	Log_{10} (relative interaction strength)	Log_{10} (relative interaction strength)	(20.7)
2e2&	0.0	+21.2	4	0.0	+42.6	(20.8)
4e4&	−21.2	0.0	3	−42.6	0.0	(20.9)
6e6&	−42.5	−21.2	2	−85.3	−42.7	(20.10)
8e8&	−63.7	−42.5	1	−128.1	−85.5	(20.11)

For each of the columns noting relative vertex strengths per r/chan, successive numbers differ by approximately −21.2474. The next item pertains.

$$Z \cdot 10^{-21.2474\ldots} \approx 6.7584 \qquad (20.12)$$

We discuss lengths at which photon-graviton series forces match maximal-% series forces

The next items pertain.

$F_{\$}(R)$ denotes the magnitude of the force associated with \$e\$& between 2 electrons separated by a distance R	(20.13)
For example, for \$=2, • The interaction is carried by 2e2& • $F_{\$}(R) = (q_e^2/4\pi\varepsilon_0)\,(1/R)^2$	(20.14)

The next items show a ratio of magnitudes of forces pertaining to interactions between 2 electrons. Here, λ_{*4} is any positive length.

$$F_2(R)\,F_4(R)\,/\,F_4(R) = F_2(R) \quad (20.15)$$
$$F_2(R)\,F_4(R)\,/\,F_4(R) = F_2(\lambda_{*4})\,(\lambda_{*4}/R)^2 \quad (20.16)$$

The left side of (20.16) is proportional to the ratio of the magnitude of the 4e24& interaction between 2 electrons to the magnitude of the 4e4& interaction between 2 electrons.

The next items show similar expressions. The first item pertains to 6e246& and 6e6&. The second item pertains to 8e2468& and 8e8&.

$$F_2(R)\,F_4(R)\,F_6(R)\,/\,F_6(R) = F_2(\lambda_{*6})\,F_4(\lambda_{*6})\,(\lambda_{*6}/R)^4 \quad (20.17)$$
$$F_2(R)\,F_4(R)\,F_6(R)\,F_8(R)\,/\,F_8(R) = F_2(\lambda_{*8})\,F_4(\lambda_{*8})\,F_6(\lambda_{*8})\,(\lambda_{*8}/R)^6 \quad (20.18)$$

The next items pertain. Here, we discuss lengths $\lambda_{*\ddot{\imath}}$ at which basic boson forces match maximal-% forces. Here, we are exploring the possibility that $\lambda_{*\ddot{\imath}} \sim \lambda_{\ddot{\imath}}$, for ï = 4, 6, and 8. [Item (6.10)]

Gss.20.2	For interactions between 2 electrons, the strengths of 4e4& and 4e24& are equal at a particle separation of roughly λ_4.	(20.19)
Gss.20.3	For interactions between 2 electrons, the strengths of 6e6& and 6e246& are equal at a particle separation of roughly λ_6.	(20.20)
Gss.20.4	For interactions between 2 electrons, the strengths of 8e8& and 8e2468& are equal at a particle separation of roughly λ_8.	(20.21)

For the moment, we assume the next item pertains.

$$\lambda_{*\ddot{\imath}} = \lambda_{\ddot{\imath}}, \text{ for } \ddot{\imath} = 4, 6, \text{ and } 8 \quad (20.22)$$

Assuming item, (20.22), the next items pertain to interactions between 2 electrons.

$$2e2\&(R) = (q_e^2/4\pi\varepsilon_0)\,(1/R)^2 \quad (20.23)$$
$$4e24\&(R) = 4e4\&(R)\,(\lambda_4/R)^2 = 2e2\&(R)\,((4/3)\beta^{12})^{-1}\,(\lambda_4/R)^2 \quad (20.24)$$
$$6e246\&(R) = 6e6\&(R)\,(\lambda_6/R)^4 = 2e2\&(R)\,((4/2)\beta^{24})^{-1}\,(\lambda_6/R)^4 \quad (20.25)$$
$$8e2468\&(R) = 8e8\&(R)\,(\lambda_8/R)^6 = 2e2\&(R)\,((4/1)\beta^{36})^{-1}\,(\lambda_8/R)^6 \quad (20.26)$$

The next items show results from using item (20.22). Relative to the strength of gravity (4e4&), entries show the log-base-10 of approximate values for the strengths of interactions between 2 electrons that are 1 m apart. Values ignore signs that would connote attraction or repulsion.

E-family member	Log$_{10}$ (magnitude of relative interaction strength), for 2 electrons, 1 m apart	(20.27)
4e4&	0.0	(20.28)
2e2&	+42.6	(20.29)
4e24&	−113.7	(20.30)
6e246&	−358.5	(20.31)
8e2468&	−691.6	(20.32)

The next items show values pertaining to hypothetical \$e. Each \$e reflects the concept of extrapolating by removing elements from % in a series for which \$e% has non-zero numbers of elements in %. For Ø denoting the null set (or, empty set), conceptually \$e denotes \$eØ. Presumably, the \$e values depend neither on the properties of 2 interacting electrons nor on the distance between the 2 interacting electrons. Values ignore signs that would connote attraction or repulsion. The value for 4e4& pertains to 2 electrons separated by 1 m.

E-family member or hypothetical member	Log$_{10}$(relative interaction strength)	(20.33)
4e4&	0.0	(20.34)
2e	112.2	(20.35)
4e	113.7	(20.36)
6e	115.2	(20.37)
8e	116.6	(20.38)

The next item pertains to items (20.35) and (20.36).

$$10^{113.7} / 10^{112.2} \approx 34.26 \approx (Z')^2 \quad (20.39)$$

We interpret the near equality of the \$e numbers as not contradicting items (20.19), (20.20), and (20.21). Perhaps, better choices of lengths at which electron-electron forces have similar strengths (better than item (20.22)) would lead both to more nearly equal values in items (20.35), (20.36), (20.37), and (20.38) and to better correlations with nature. The next items suggest possibilities.

$$\lambda_{*\ddot{\imath}} \approx \acute{o}(\ddot{\imath})\ (Z')^{(\ddot{\imath}-2)/2}\ \lambda_{\ddot{\imath}}\text{, for } \ddot{\imath} = 2, 4, 6, \text{ and } 8 \text{ and with ...} \quad (20.40)$$

$$\acute{o}(2) = 1 \quad (20.41)$$

$$\acute{o}(4) = 1$$

$$\acute{o}(6) \approx 0.9428$$

$$\acute{o}(8) \approx 0.767$$

Assuming item (20.40) pertains, the next items improve upon items including and following item (20.33).

E-family member or hypothetical member	Log$_{10}$(relative interaction strength)	(20.42)
4e4&	0.0	(20.43)
2e	112.2	(20.44)
4e	112.2	(20.45)
6e	112.2	(20.46)
8e	112.2	(20.47)

Assuming item (20.40) pertains, the next items improve upon items including and following item (20.27).

E-family member	$\log_{10}$ (magnitude of relative interaction strength), for 2 electrons, 1 m apart	(20.48)
4e4&	0.0	(20.49)
2e2&	+42.6	(20.50)
4e24&	−112.2	(20.51)
6e246&	−352.4	(20.52)
8e2468&	−678.5	(20.53)

We discuss interactions for other than just electrons and positrons

The next item pertains.

Vertex strengths scale per particle properties (20.54)

The next items pertain. These items combine item (6.23) and following items, item (20.2) and following items, and item (20.54). The subscript e denotes electron.

Interaction	R/chans	Object property	Symbol for object property	Relative vertex strength per r/chan	(20.55)
2e2&	4	PROPE2	Q	$\beta^{-0} Q / \lvert q_e \rvert$	(20.56)
4e4&	3	PROPE4	m	$\beta^{-6} m / m_e$	(20.57)
6e6&	2	PROPE6	m_{mm}	$\beta^{-12} m_{mm} / m_{mm,e}$	(20.58)
8e8&	1	PROPE8	PROPE8	$\beta^{-18} \lvert(\text{PROPE8})\rvert / \lvert(\text{PROPE8})_e\rvert$	(20.59)

with ... (20.60)

- $m_{mm,e} = g_s\hbar/2 = \hbar$

The next items extend items starting with item (20.55). Here, for each property ύ, ύ(1) denotes that property for object 1. Here, for each property ύ, ύ(2) denotes that property for object 2. These items pertain to the magnitudes of the various $e%&. We do not discuss signs.

If % contains the following symbol ...	... multiply the 2-electron result by ...	(20.61)
2	$Q(1)\ Q(2) / (q_e)^2$	(20.62)
4	$(\text{PROPE4})(1)\ (\text{PROPE4})(2) / (m_e)^2$	(20.63)
6	$m_{mm}(1)\ m_{mm}(2) / (m_{mm,e})^2$	(20.64)
8	$(\text{PROPE8})(1)\ (\text{PROPE8})(2) / ((\text{PROPE8})_e)^2$	(20.65)

We discuss a possible example of a perturbation technique

We explore a possibility that a perturbation-like technique applies. In particular, we note a possible approximation for α, the fine-structure constant. People might correlate this work with the $e2& series of e-family bosons. [Items following item (8.72)]

The next items define the approximation.

$$\alpha \approx \Sigma \text{ (r/chan ratio) } \kappa^{\gamma} (2\pi)^{\gamma'} \quad (20.66)$$

An r/chan ratio denotes the ratio of r/chans for \$ (in \$e%&) to r/chans for \$=2 (in 2e2&) (20.67)

$$\kappa = 2 \quad (20.68)$$

Here, we choose γ for κ^{γ} to match series that feature powers of β. We choose $\kappa = 2$ to correlate with the opening (term by term) of oscillator pairs. We choose γ' to obtain an approximate result. The next items show numbers.

R/chan ratio	γ in κ^{γ}	γ' in $(2\pi)^{\gamma'}$	Single term	Σ = cumulative sum of terms	$(\Sigma - \alpha) / \alpha$	(20.69)
3/4	−12	2	7.22871×10^{-3}	7.2287×10^{-3}	-9.4059×10^{-3}	(20.70)
2/4	−24	4	4.64483×10^{-5}	7.2752×10^{-3}	-3.0408×10^{-3}	(20.71)
1/4	−36	8	8.83688×10^{-6}	7.2840×10^{-3}	-1.8299×10^{-3}	(20.72)

We suggest research

STR.20.1 Develop theory sufficient to predict choices - attraction, repulsion, or neither - for each e-family interaction between 2 particles.

STR.20.2 Predict strengths and directions (attraction or repulsion) for e-family forces other than 2e2& and 4e4&.

STR.20.3 To what extent do neutrinos interact with \$e%&-for-which-2∊% bosons based on, for example, neutrinos being transformed into virtual pairs, each consisting of a charged lepton and a 2w2 or 2w3?

STR.20.4 Estimate the Hubble constant.

STR.20.5 Develop a suitable IOM perturbation theory (possibly based on something like Feynman diagrams) for e-family interactions.

STR.20.6 Use such an IOM perturbation theory [STR.20.5] to estimate magnetic-moment anomalies. [Items (6.44) and (6.45)]

Section 21 Baryonic matter, dark matter, and dark-energy stuff

Abs.21.1 We note and interpret observations regarding effects of dark matter and dark-energy stuff.
Abs.21.2 We discuss 2 possible types of DMF (dark-matter basic fermions) - S=3/2 basic fermions and siblings of BMCF (baryonic-matter charged fermions).
Abs.21.3 We discuss 2 possible types of DEF (dark-energy basic fermions) - S=7/2 basic fermions and peers of DM`N`BMC`F (DMF + NF (neutrinos correlating with DMF + BMCF) + BMCF).
Abs.21.4 We discuss 2 possible types of BDEF (beyond-dark-energy basic fermions) - S= 9/2 basic fermions and a peer of NBDEF (DEF + DM`N`BMC`F).

Context

We note observations about densities of the universe

People observe characteristics of galaxies and galactic clusters that people say correlate with more mass than people think baryonic matter provides. People use the term dark matter to name stuff that correlates with the needed mass.

People discuss non-uniformities in CMB (cosmic microwave background) radiation. People interpret some of the non-uniformities as correlating with the existence of things people call dark matter and dark energy.

People estimate ratios of contributions to the density of the universe of baryonic matter, dark matter, and dark energy. People base estimates on data about CMB. Traditional physics considers that observed CMB photons have existed since near the time of the big bang. Physics considers that before $10^{5.6}$ years after the big bang, such photons interacted significantly with ionized plasma. Around $10^{5.6}$ years after the big bang, the plasma ceased to be ionized. Much CMB radiation travels today.

Data suggest late-time (or, secondary) anisotropies. Late-time refers to any time after the above-mentioned plasma ceased to be ionized. Relevant processes for generating non-uniformities in CMB may feature photon scattering by free electrons (Thompson scattering), scattering by clouds of high-energy electrons (Compton scattering by hot electron gases), and frequency-shifting because of changing gravitational fields (integrated Sachs-Wolfe effect). [Ref.21.1]

The next items note interpretations of observations pertaining to recent times in the history of the universe. [Ref.21.2]

A ratio of densities for baryonic matter and dark matter is $\sim 1 : 5.5$ (21.1)
A ratio of densities for baryonic matter and dark energy is $\sim 1 : 13.9$ (21.2)
A ratio of densities for non-dark-energy (baryonic matter and dark matter) and dark energy is $\sim 1 : 2.2$ (21.3)

As far as we know, people might characterize traditional concepts of details (such as properties of basic particles) of dark matter and dark energy (dark-energy stuff) as uncertain or vague.

We anticipate offering choices regarding the nature of dark matter and dark-energy stuff

We discuss possibilities for dark matter and for dark-energy stuff.

Core

We provide working definitions for baryonic-matter, dark-matter, and dark-energy fermions

The next items correlate with a means to frame a discussion regarding baryonic matter, dark matter, and dark energy. Above, we do not use the term dark energy to refer to a force governing the rate of expansion of the universe. Here, we use the term dark-energy fermions to refer to some fermions contributing to the density of the universe. Here, we show how 1 hypothetical type of basic fermions - S=5/2 basic fermions - might pertain. But, below, we continue to assume that S=5/2 basic fermions do not pertain.

Terminology for fermions: • Acronym • Concept	Concept: The basic fermions do not interact directly with baryonic-matter charged basic fermions via ...	Examples	(21.4)
• BDEF • Beyond-dark-energy basic fermions	• 8e8& • forces listed below	• S=9/2 basic fermions	(21.5)
• DET2F • Dark-energy-type-2 basic fermions	• 6e6& • forces listed below	• S=7/2 basic fermions	(21.6)
• DET1F • Dark-energy-type-1 basic fermions	• 4e4& • the force listed below	• (S=5/2 basic fermions)	(21.7)
• DMF • Dark-matter basic fermions	• 2e2&	• S=3/2 basic fermions	(21.8)
• NF • Neutrinos	• 2e2&	• Neutrinos	(21.9)
• BMCF • Baryonic-matter charged basic fermions	• (none)	• S=1/2 charged basic fermions	(21.10)

The next items define additional acronyms.

• Acronym • Concept	Definition	(21.11)
• DM\`N\`BMC\`F • Dark-matter plus neutrino plus baryonic-matter-charged basic fermions	DMF + NF + BMCF	(21.12)
• DEF • Dark-energy basic fermions	DET2F + DET1F	(21.13)
• NBDEF • Not-beyond-dark-energy basic fermions	DET2F + DET1S + DM\`N\`BMC\`F	(21.14)

The next item discusses concepts regarding neutrinos. [Items (21.4), (21.8), (21.9), and (21.10)]

Neutrino-photon interactions are indirect, for example via a neutrino's creating a virtual pair of W bosons (21.15)

We discuss aspects related to the applicability of the working definitions

We think the above working definitions pertain adequately well for the late-time era and for some time before that era. During yet earlier times, o- and w-family-mediated interactions presumably provided contributions significant compared to e-family-intermediated interactions. Presumably, the end of this yet-earlier-time period correlates somewhat with the formation of individual nucleons.

We discuss possible implications of symmetries

The next item might correlate with possibly relevant symmetries. Here, people might say that the 8 instances of 4e4& correlate with an E3-through-E1 e-family symmetry (SU(3)). [Item (12.11)] Here, people might say that 6 instances of the s-family correlate with an E3-through-E1 s-family symmetry (SU(2)×U(1)). [Item (12.23)] Or, here, people might say that 6 instances of 2e2& correlate with a relationship between the SU(3) symmetry and an SU(7) symmetry. [Items including and following item (13.4)] And, people might say that 6 instances of either the s-family or of 2e2& correlate with 6 instances of all S=1 basic bosons.

- 8 instances of 4e4& exist (21.16)
- For each instance of 4e4& particles, 6 instances of S=1 basic bosons exist

People might say that the next items correlate with item (21.16).

- The symmetry correlates with 8 (=2×2×2) ways to align each of P4-and-P5 (2 ways), P6-and-P7 (2 ways), and P8-and-P9 (2 ways) with P10-and-P11 (or, to align P4 through P9 with the QP-like 10-axis and QP-like 11-axis in energy-momentum space) (21.17)
- The symmetry correlates with 6 (=3! or =3×2) ways to align P1, P2, and P3 with the QP-like 1-axis, 2-axis, and 3-axis in energy-momentum space
- 2 instances of 8e8& particles exist (21.18)
- For each instance of 8e8& particles, 2 instances of 6e6& particles exist
- For each instance of 6e6& particles, 2 instances of 4e4& particles exist
- For each of the 8 instances of 4e4& particles, 6 instances of (21.19)
  - BMCF + S≤1 basic bosons exist

We provide possibilities for beyond-dark-energy basic fermions

The next items list possibilities for beyond-dark-energy basic fermions.

S=9/2 basic fermions [Item (21.5)] (21.20)
4 peers of DM`N`BMC`F, each based on 1 instance of 4e4& (gravity) that does not interact directly with NBDEF [Item (21.19)] (21.21)
Other (21.22)

We do not discuss item (21.22) further.

We do not rule out either S=9/2 basic fermions or 4 peers of DM`N`BMC`F as candidates for beyond-dark-energy basic fermions

The next item pertains.

Gss.21.1	2 not-mutually-exclusive candidates exist for BDEF. 1 candidate is S=9/2 basic fermions. The other candidate is 4 peers of DM`N`BMC`F.	(21.23)

Except possibly for an asymmetry early in the big bang, people might say that BDEF would not impact characteristics of CMB.

We provide possibilities for dark-energy basic fermions

The next items list possibilities for dark-energy basic fermions.

S=7/2 basic fermions [Item (21.6)]	(21.24)
3 peers of DM`N`BMC`F, each based on 1 instance of 4e4& (gravity) that does not interact directly with BDEF or with DM`N`BMC`F [Item (21.19)]	(21.25)
Other	(21.26)

We do not discuss item (21.26) further.
Of the 3 peers of DM`N`BMC`F, 2 peers correlate with DET2F and 1 peer correlates with DET1F.

We discuss delays in DEF impacting CMB

Analyses of CMB data determine, in effect, aspects of clumping. We use the term clumping to denote non-homogeneity in the distribution of energy or matter. BMCF clumps correlate with atomic nuclei, atoms, planets, stars, solar systems, galaxies (which can include significant amounts of DMF), galactic clusters, clouds of electrons, and so forth. Possibly, DEF (or, DET2F + DET1F) undergoes clumping. Presumably, DMF undergoes clumping, somewhat on its own and somewhat in coordination with BMCF. (Some objects, such as galaxies, correlate with DMF clumps.)

For DEF to have fully impacted CMB, activity the next items list may need to occur. (People might downplay references to NF. Perhaps, effects of NF clumping have little indirect or direct impact on CMB.)

DET2F has clumped	(21.27)
If DET1F exists, 8e%& interactions have induced clumping in DET1F based on clumping in DET2F	(21.28)
8e%& interactions and 6e%& interactions have induced clumping in DMF and NF based on clumping in DET2F and (if it exists) DET1F	(21.29)
8e%& interactions, 6e%& interactions, and 4e%& interactions have induced clumping in BMCF based on clumping in DET2F, DET1F (if it exists), DMF, and NF	(21.30)
CMB photons (some of the 2e2& associated with BMCF) have reacted to the following • Clumping of BMCF objects • Effects of clumping on the gravitational field (and other fields) germane to CMB photons	(21.31)

Possibly, people should also consider early-universe effects of BDEF clumping. People might say that, here, early-universe incudes times until effects of 10o% bosons become not significant compared to effects of other bosons.

We do not rule out either S=7/2 basic fermions or 3 peers of DM`N`BMC`F as candidates for dark-energy basic fermions

The next item pertains.

Gss.21.2	2 not-mutually-exclusive candidates exist for DEF (or, DET2F + DET1F). 1 candidate is S=7/2 basic fermions. The other candidate is 3 peers of DM`N`BMC`F.	(21.32)

Based on items including item (21.31), neither item (21.2) not item (21.3) seems to rule out either candidate. (Perhaps, data of which we are not aware could rule out a candidate.)

We provide possibilities for dark-matter basic fermions

S=3/2 basic fermions provide 1 candidate for dark-matter basic fermions.

Above, we discuss a symmetry that might correlate with 6 instances of BMCF for each instance of 4e4&. [Item (21.19)]

The next item pertains. [Item (18.55)]

Neutrinos correlate with an 8-fold (or, SU(3)-related) symmetry that people might interpret as indicating that 1 set of neutrinos correlates with DMF + BMCF (or, that there is no relevant difference between baryonic-matter neutrinos and dark-matter neutrinos)	(21.33)

We use the term ensemble to denote an instance of BMCF particles.

The next item extends item (21.19).

Possibly, • For each instance of 4e4& particles, 6 ensembles exist	(21.34)

The next items list possibilities for dark-matter basic fermions.

S=3/2 basic fermions	(21.35)
5 ensembles other than the BMCF ensemble	(21.36)
Other	(21.37)

We do not further discuss item (21.37).

We do not rule out either S=3/2 basic fermions or 5 siblings of BMCF as candidates for dark-matter basic fermions

The next item pertains.

Gss.21.3 2 not-mutually-exclusive candidates exist for DMF basic fermions. 1 candidate is S=3/2 basic fermions. The other candidate is 5 ensembles that are siblings of the BMCF ensemble. (21.38)

Comments

We discuss o-family-mediated interactions between non-BMCF and BMCF

To the extent S≥3/2 basic fermions exist, the next items indicate possibilities for o-family-mediated interactions between non-BMCF and BMCF.

- 3qb(...) and 3qa(...) (which we classify above as DMF) possibly interact with BMCF quarks via $o3 bosons and via $o2 bosons [Items following item (11.55)] (21.39)
- 7qb(...) and 7qa(...) (which we classify above as DET2F) possibly interact with BMCF quarks via $o3 bosons and via $o2 bosons [Items following item (11.74)] (21.40)
- 9q(2,8), 9q(3,8), 9q(2,9), and 9q(3,9) (which we classify above as BDEF) possibly interact with BMCF quarks via $o3 bosons and via $o2 bosons [Items following item (11.80)] (21.41)

We discuss masses for S≥3/2 basic fermions

To the extent S≥3/2 basic fermions exist, perhaps the concept of spin/mass ratio [Item (6.15) and also items (6.51) and (6.52)] provides insight into masses for S≥3/2 basic fermions. For example, perhaps the BMCF-centric mass for the generation-1 3qï(ó)-related particle (for ï being b and/or a [Items following item (11.55)]) is 3 (=(3/2)/(1/2)) times the mass of a corresponding BMCF up or BMCF down.

We do not further discuss possibilities for masses for S≥3/2 basic fermions.

We discuss the range of applicability of the concept of ensemble

To the extent DMF includes 5 ensembles (and not just 1 ensemble), presumably the next item pertains.

- DET1F includes 6 ensembles and 1 peer of NF (21.42)
- DET2F includes 12 ensembles and 2 peers of NF
- BDEF includes 24 ensembles and 4 peers of NF

Thus, to the extent DMF includes 5 ensembles (and not just 1 ensemble), people might say that the universe includes 48 (=6+6+12+24) ensembles and 8 (=1+1+2+4) sets of neutrinos.

We discuss topics related to properties of the possible DMF S=1/2 basic fermions

Here, we assume DMF includes 5 ensembles. The next items pertain.

Possibly, properties for a basic fermion in a DMF ensemble equal properties for the counterpart basic fermion in the BMCF ensemble ... (21.43)

- As measured by interactions with the BMCF ensemble

... and / or ...

- As measured via means that are not centric to the BMCF ensemble

Possibly, properties for a basic fermion in a DMF ensemble differ from properties for the counterpart basic fermion in the BMCF ensemble ... (21.44)
- As measured by interactions with the BMCF ensemble

... and / or ...
- As measured via means that are not centric to the BMCF ensemble

Item (21.44) provides for 2 possible ways in which properties measured for a DMF-ensemble basic fermion might differ from properties for the corresponding BMCF-ensemble basic fermion.

To the extent either such difference pertains, we think, for charged S=1/2 basic fermions, work in Section 18 pertains, but with ratios of charges to masses based on 3≤M"≤18. Items following item (18.58) would pertain. ("DMF" group 0 correlates with the BMCF ensemble. "DMF" groups 1 through 5 correlates with DMF ensembles.) For example, for each difference, 1 (but not both) of the next 2 items could describe charges and masses for least-mass dark-matter charged leptons.

$|q| = |q_e|$, for DMF ï, with 1≤ï≤5 (21.45)
$m = \beta^{ï} m_e$, for DMF ï, with 1≤ï≤5

$|q| = \beta^{-ï} |q_e|$, for DMF ï, with 1≤ï≤5 (21.46)
$m = m_e$, for DMF ï, with 1≤ï≤5

Above, we discuss the possibility that masses for all neutrinos that interact directly via 4e4& with DMF + BMCF correlate with masses we calculate for m(ó,0), with ó = −6, −4, and −3. This possibility might correlate with item (21.46).

The next items pertain.

- 1 instance of gravity (4e4&) pertains to and spans DM`N`BMC`F (21.47)
- 1 instance of neutrinos pertains to DM`N`BMC`F (21.48)
- Masses of DMF S=1/2 basic fermions equal masses for counterpart BMCF S=1/2 basic fermions (21.49)
  - People might associate the term anti-tauon with the DMF 1 counterpart to a BMCF electron [Item (18.62)]

To the extent items (21.44) and (21.46) pertain, people might say that the next items pertain.

- In the reference frame of DMF and in the reference frame of BMCF (item (21.44)), item (21.46) pertains (21.50)
- For e-family (or, \$e%&) mediated interactions other than \$e%&-mediated interactions for which 2∈%, ... (21.51)
  - strengths of DMF interactions = strengths of BMCF interactions = strengths of DMF-BMCF interactions
- With the definition (based on mass) of counterpart particles, for charges q ... (21.52)
  - |q(DMF S=1/2 basic fermion)| = $\beta^{-ï}$ |q(counterpart BMCF S=1/2 basic fermion)|
    - [Item (21.46)]

To the extent items (21.44) and (21.46) pertain, people might say the next items pertain.

- DMF (atomic) nuclei span a wider range of atomic numbers than do BMCF nuclei (21.53)
  - For DMF nuclei, charge-based repulsion of baryons is less than the similar repulsion in BMCF nuclei
- DMF atoms or ions are larger than BMCF atoms or ions (21.54)
  - DMF attraction between charged leptons and nuclei is less than counterpart BMCF attraction between charged leptons and nuclei
- Possibly, most DMF charged leptons are not components of DMF atoms or ions (21.55)
- Possibly, much of DMF exists in forms such as the following (21.56)
  - Fully ionized plasma
  - Analogs to BMCF white dwarf stars
    - Or, components of DMF-BMCF white dwarf stars
  - Analogs to BMCF neutron stars
    - Or, components of DMF-BMCF neutron stars
  - Analogs to BMCF black holes
    - Or, components of DMF-BMCF black holes

We discuss basic masses for possible S=1/2 basic fermions belonging to DET1F, DET2F, and BDEF

To the extent DMF+BMCF includes 6 ensembles (and not just 1 ensemble), DET1F includes 6 ensembles.

DMF+BMCF cannot measure DET1F, DET2F, or BDEF basic fermion masses via interactions mediated solely by 2e2& or 4e%& e-family members.

Perhaps, DET1F S=1/2 basic fermion masses equal masses for basic fermions in the 6 DMF+BMCF ensembles.

Or, perhaps masses vary by ensemble in a manner paralleling the variation of charge by ensemble that item (21.46) shows for charges. (Compare the DEF group column in items including and following item (18.78) with the DMS group column in items including and following item (18.58).) Here, for example, DET1F S=1/2 basic fermion masses could be much smaller than masses for basic fermions in the 6 DMF+BMCF ensembles.

We continue to discuss cases for which the number of NBDEF ensembles is 24

Assuming that the number of NBDEF ensembles is 24, the next items apply.

Relative to any 1 ensemble, ... (21.57)
- 5 ensembles of dark-matter basic fermions exist
- 18 ensembles of dark-energy basic fermions exist
- 24 ensembles of beyond-dark-energy basic fermions exist

Dark-matter correlates with a reciprocal relationship between 2 ensembles (21.58)
- The 2 ensembles share an instance of 4e4& forces

Dark-energy correlates with a reciprocal relationship between 2 ensembles (21.59)
- The 2 ensembles do not share an instance of 4e%& forces
- The 2 ensembles share an instance of 8e%& forces

Beyond-dark-energy correlates with a reciprocal relationship between 2 ensembles (21.60)
- The 2 ensembles do not share an instance of any $e%& forces

We suggest research

SOR.21.1 Find or rule out (to some confidence level) evidence (other than currently assumed evidence) of effects on baryonic matter of dark matter.

SOR.21.2 Detect (or infer) or rule out (to some confidence level) that 4e24% mediates interactions between BMCF and DMF. (Here, we have in mind that 4e4& mediates interactions between BMCF and DMF and that 2e2& does not mediate direct interactions between BMCF and DMF.)

SOR.21.3 Measure or infer masses for (or rule out (to some confidence level) existence of) S≥3/2 basic fermions.

SOR.21.4 Determine the extents to which fields related to various 4e%&, 6e%&, and 8e%& bosons have impacted CMB radiation.

SOR.21.5 Determine or rule out (to some confidence level) the existence of more than 1 ensemble.

SOR.21.6 Determine the extents to which fermion properties vary by ensemble.

SOR.21.7 Infer w-, h-, and o-family interaction strengths and/or big-bang phenomena related to w-, h-, and o-family-mediated interactions.

SOR.21.8 Detect (or infer) or rule out (to some confidence level) effects on NBDEF of BDEF.

SOR.21.9 Determine the extent to which observations (for example, about the uniformity of distribution of NBDEF) cannot be explained by traditional physics and/or work in this paper directly related to NBDEF. Estimate the extent to which people might attribute any such unexplained phenomena to the presence of BDEF (for example, via effects early in the big bang).

SOR.21.10 Determine the extent to which (DMF, DET1F, DET2F, and/or BDEF) ensembles contribute to actual densities of the universe. (Here we differentiate actual from inferred {for example, inferred via CMB data}.)

STR.21.1 How best should people try to directly detect matter not associated with baryonic matter?

STR.21.2 Develop models people can use to estimate masses for S≥3/2 basic fermions.

STR.21.3 Estimate ratios of densities of the universe for baryonic matter, dark matter, dark-energy stuff, and beyond-dark-energy stuff, assuming that S≥3/2 basic fermions provide all DMF fermions, all DET2F fermions, and all BDEF fermions.

STR.21.4 To what extent might people benefit by considering the possibility that an 8-fold symmetry among 8e8&, 6e6&, and 4e4& correlates with the possibility that nature does not exhibit negative values of each of PROPE8, PROPE6, and PROPE4? [Item (21.17)]

STR.21.5 Review observational data and calculations of the density of baryonic matter (equivalent to a mass density of approximately 9.9×10^{-30} g/cm^3, which is similar to mass density of 5.9 protons per cubic meter) to find a basis for determining the extent to which S≥3/2 fermions might pertain. [Ref.21.3]

STR.21.6 For each of the various eras (ZEú, per Section 16), which interactions most affect neutrinos?

We list references

Ref.21.1 J. Beringer et al. (Particle Data Group), *Phys. Rev. D86*, 010001 (2012). (http://pdg.lbl.gov/2012/reviews/rpp2012-rev-cosmic-microwave-background.pdf)

Ref.21.2 Mark Peplow, Planck telescope peers into primordial universe, *Nature News*, Nature Publishing Group, March 21, 2013. (http://www.nature.com/news/planck-telescope-peers-into-primordial-universe-1.12658)

Ref.21.3 Wilkinson Microwave Anisotropy Probe, http://wmap.gsfc.nasa.gov/universe/WMAP_Universe.pdf

Section 22 Matter/antimatter imbalance and CPT-related symmetries

Abs.22.1 Lasing of o-family particles provided key effects leading to matter/antimatter imbalance.
Abs.22.2 E-family coherences provide for phenomena people attribute to axions.
Abs.22.3 O-family particles provide for CPT-related symmetry violations.
Abs.22.4 O-family particles with S≥2 close gaps between magnitudes of violations people estimate via the Standard Model and magnitudes people measure.

Context

We discuss various asymmetries

People speculate as to why baryonic matter has much more matter (such as electrons and protons) than antimatter (such as positrons and antiprotons). People speculate as to sources of observed symmetry violations, for example with respect to P (parity) or CP (charge and parity). People say that the Standard Model can account for some P violation and CP violation. People say that the Standard Model does not account for observed amounts of P violation and CP violation.

We anticipate providing explanations regarding asymmetries

Here, we provide possible explanations. We discuss o- and e-family phenomena with which the Standard Model does not correlate.

Core

We discuss matter/anti-matter imbalance

The next items show possible examples of interaction vertices. Such interactions would convert quarks into anti-quarks or anti-quarks into quarks. Depending on which IOM model is appropriate, values of $ = 4, 6, 8, and 10 may also pertain. [Section 17]

Interaction vertices		|Q'|, for $o%	
			(22.1)
up + $o2 → anti-down	for $ = 2	1/3	(22.2)
anti-up + $o3 → down	for $ = 2	1/3	(22.3)

The next item shows a reaction that would convert a positron and the basic q-family components for an anti-proton into an electron and the basic q-family components for a proton.

1 Positron + 2 Anti-up + 1 Anti-down → 1 Electron + 2 Up + 1 Down (22.4)

The next items show reactions that would net to item (22.4). Here, each entry shows a value of Q'. Here, a fermion name with its first letter capitalized represents a fermion named in item (22.4). A fermion name with its first letter in lower case represents a fermion that the reactions create and then destroy. A +1 boson corresponds to a W^+. A −1 boson corresponds to a W^-. A +1/3 boson corresponds to a $o3 having Q'=+1/3.

This fermion	emits or absorbs this boson	and becomes this fermion.	This fermion	absorbs or emits this boson	and becomes this fermion.	(22.5)
−2/3 Anti-up	absorbs +1	+1/3 anti-down	+1 Positron	emits +1	0 neutrino	(22.6)
−2/3 Anti-up	emits −1	+1/3 anti-down	0 neutrino	absorbs −1	−1 Electron	(22.7)
+1/3 Anti-down	absorbs +1/3	+2/3 up	+2/3 up	emits +1	−1/3 Down	(22.8)
+1/3 anti-down	absorbs +1/3	+2/3 Up				(22.9)
+1/3 anti-down	absorbs +1/3	+2/3 Up				(22.10)

These reactions seem to correlate with reactions such as the next item shows. The next item correlates with the minimum number (3) of bosons item (17.34) shows for 2o3.

$$? + W^{+} = ? + 2w3 \rightarrow 3\ 2o3 + ?' \qquad (22.11)$$

Presumably, during part of the big bang, baryonic matter was sufficiently dense that many basic q-family particles interacted via $o%. A non-uniformity (in, for example, spatial distributions of quarks) or an instability could have led to lasing by 2o3 bosons. Possibly, the ? and ?' in item (22.11) could include e-family particles, which also could have lased.

The next items list imbalances such reactions would create.

Number of Q'=−1 leptons ≫ number of Q'=+1 (anti-)leptons (22.12)
Number of Q'=+2/3 quarks ≫ number of Q'=−2/3 (anti-)quarks (22.13)
Number of Q'=−1/3 quarks ≫ number of Q'=+1/3 (anti-)quarks (22.14)

We discuss an example of P-symmetry violation

The next item pertains.

Gss.22.1 People consider interactions that convert anti-quarks into quarks (or vice-versa) to violate P symmetry. (22.15)

We base the next item on reactions such as items following item (22.1) indicate.

0-family bosons ($o3 and $o2, for at least $=2) mediate interactions that exhibit P violation. (22.16)

We discuss CP violation and the possibility for axions

People say that CP violations correlating with the Standard Model may be too small to correlate with observed magnitudes of CP violations. People note a concept of axions. People suggest axions might provide for some CP violation. People suggest that axions have some (small) mass and may decay into photons. We suggest the following.

Gss.22.2 Phenomena people associate with axions exist. People can associate such phenomena with e-family coherences. (22.17)

E-family member 4e24& becomes a candidate for participating in such phenomena. Here, a relevant vertex could involve the 4e4& (graviton) component of a 4e24% and could leave the 2e2& (photon) component as a product. Here, the using of the graviton component could correspond to traditional thoughts that people express as non-zero mass for axions. Possibly, any \$e%& for which 2∈% and for which % has more than 2 elements contributes to phenomena people associate with axions.

We reinterpret traditional P-, C-, and T-symmetries and related violations

The next items show maximum #E and maximum #P for IOM(11,11).

Families	Maximum #E for IOM(11,11)	Maximum #P for IOM(11,11)	(22.18)
e- and s-	7	9	(22.19)
w-, h-, and o-	11	11	(22.20)
l- and q-	11	11	(22.21)

The next items show similar maximums for IOM(3,3), which we think correlates with the Standard Model.

Families	Maximum #E for IOM(3,3)	Maximum #P for IOM(3,3)	(22.22)
e- and s-	3	3	(22.23)
w-, h-, and o-	3	3	(22.24)
l- and q-	3	3	(22.25)

People traditionally discuss P-, C-, and T-symmetries based on physics that items following item (22.22) model.

The next item describes aspects of those symmetries. Here, the $p_{ó}$ refer to components of a 4-momentum 4-vector, with p_0 corresponding to the energy-related component.

Family	P-symmetry exchanges	T-symmetry exchanges	(22.26)
e-	$p_{ú} \leftrightarrow -p_{ú}$, $1 \leq ú \leq 3$	$p_0 \leftrightarrow -p_0$	(22.27)

IOM provide that n_{P1} aligns (for e-family basic bosons) with motion. The next items pertain for IOM(3,3).

Families	C-symmetry exchanges	P-symmetry exchanges	T-symmetry exchanges	(22.28)
e- and s-	E3 ↔ E2 P2 ↔ P3	P2 ↔ P3	E3 ↔ E2	(22.29)
w- and o-	E3 ↔ E2 P2 ↔ P3	P2 ↔ P3	E3 ↔ E2	(22.30)
l- and q-	E3 ↔ E2 P2 ↔ P3	P2 ↔ P3	E3 ↔ E2	(22.31)

Items following item (22.28) exhibit CPT symmetry. Also, the h-family exhibits CPT symmetry. (Only oscillators E1 and P1 pertain to relevant aspects of the h-family.) People state that CPT symmetry applies to Standard Model physics.

Exchanges the next items list possibly pertain for IOM(11,11) for the e-, o-, and q- families.

C-symmetry exchanges	P-symmetry exchanges	T-symmetry exchanges	(22.32)
E11 ↔ E10 E9 ↔ E8 E7 ↔ E6 E5 ↔ E4 E3 ↔ E2 P2 ↔ P3 P4 ↔ P5 P6 ↔ P7 P8 ↔ P9 P10 ↔ P11	P2 ↔ P3 P4 ↔ P5 P6 ↔ P7 P8 ↔ P9 P10 ↔ P11	E11 ↔ E10 E9 ↔ E8 E7 ↔ E6 E5 ↔ E4 E3 ↔ E2	(22.33)

For each of the e-, o-, and q- families, some exchanges beyond the exchanges items following item (22.28) list pertain. For example, P4 ↔ P5, P6 ↔ P7, and P8 ↔ P9 pertain for the e-family. For example, E9 ↔ E8, E7 ↔ E6, and E5 ↔ E4 pertain for the o-family. Per item (22.33), each exchange exhibits CPT symmetry.

The next item pertains.

Gss.22.3 Differences between ύ-symmetry for IOM(11,11) and ύ-symmetry for IOM(3,3) correlate with sizes of ύ-violations people do not associate with Standard Model physics. Here, ύ can be (at least) C, P, or T. (22.34)

Comments

We suggest research

SOR.22.1 Detect or rule out (to some confidence level) that 2o3 and 2o2 bosons can covert a q-family S=1/2 fermion from anti-quark to quark and vice-versa.

SOR.22.2 Determine the extents to which w- and o-family-mediated interactions and/or c-family-mediated interactions account for observed P violations, CP violations, or other such violations.

STR.22.1 To what extent do e-family-coherence-mediated interactions correlate in strength with measured CP violations?

STR.22.2 To what extent do \$0%-mediated interactions correlate in strength with measured C, P, and T violations?

STR.22.3 To what extent do some o-family-mediated interactions correlate in strength with C, P, and T violations for which the Standard Model can account?

Section 23 Other topics

Abs.23.1 $\Omega<0$ correlates with an inability for people to observe particles as free-ranging and with particles' exhibiting quantum QP-like behavior.

Abs.23.2 Perhaps, aspects of the uncertainty principle correlate with 2e2& being at least 2 of a basic boson, maximal-% boson, and \$e2&-series boson.

Abs.23.3 People may derive from IOM useful insight about handedness/chirality.

Abs.23.4 Maximal-% e-family forces pertain regarding black holes and may produce quasars.

Abs.23.5 IOM may correlate with interactions people say (hypothetical) leptoquarks might mediate.

Context

We note that work above leaves some topics not completely addressed

Work above does not discuss various possibly relevant topics and possibilities.

We note that we can discuss some not completely addressed issues

We anticipate making remarks about some topics and possibilities.

Core

We discuss the possibility of space-like (or, faster-than-light-speed) behavior

The next items pertain. [Item (15.5)]

Gss.23.1 In the sense of traditional quantum perturbation models for interactions between elementary particles, people might consider that q- and o-family particles traverse space-like trajectories between vertices. (23.1)

Gss.23.2 Item (23.1) correlates with an inability to observe free-ranging instances of $\Omega<0$ basic particles. (23.2)

We explore uncertainty related to solutions

We use a QM-type-CS approach. The next item follows from item (4.24).

$$\xi = (\xi_0/2)\,(\,\eta^2\langle p_r^2\rangle + \eta^{-2}\langle r^2\rangle\,), \text{ in which } \quad (23.3)$$

$\langle ύ \rangle$ denotes the expected value of ύ

The next item follows from item (4.35).

$$D + 2\nu = \eta^2\langle p_r^2\rangle + \eta^{-2}\langle r^2\rangle \quad (23.4)$$

$\eta^{-2}\langle r^2\rangle$ does not depend on η. Similarly, $\eta^2\langle p_r^2\rangle$ does not depend on η. The 2 terms contribute equally. The next items pertain for $\xi_0 \neq 0$.

$\langle p_r^2\rangle \times \langle r^2\rangle$ does not vary with changes in η^2, for $\eta^2>0$	(23.5)
For edge solutions, people might consider that $\langle r^2\rangle = 0$	(23.6)

We discuss a possible way to derive an uncertainty-like equation for w-family basic bosons

The next items sketch deriving an uncertainty-like equation for w-family bosons.

Follow steps correlating with items (14.1) through (14.28) (23.7)
Set $\upsilon_0 = -\xi_0$ (23.8)
- Compare with item (14.29)

Note the following (compare with item (14.33)) (23.9)
- $\ddot{\iota}^2 E^2 + \ddot{\iota}^{-2} t^2 + \eta^2 c^2P^2 + \eta^{-2} c^2x^2 \propto$ (components from 2-dimensional QE-like terms and components from 2-dimensional QP-like terms)

We discuss basis states and a possible correlation with uncertainty

Work in this paper features basis states. People might say that no traditional uncertainty applies regarding kinematics for basis states. (People might say the same thing regarding plane-wave solutions people use within traditional quantum mechanics.) In traditional plane-wave kinematics, the uncertainty principle and its use of the constant $\hbar$ arise regarding the superposition of basis-state wave functions. Presumably, the same happens regarding IOM.

We explore that possibility that IOM produce, from a quantum basis and for 1 basis state, the notion that $\hbar$ pertains to uncertainties. Perhaps people would say that items including and following item (23.7) point toward a correlation between $\hbar$ and basis states.

The next item pertains.

Gss.23.3	The confluence of at least 2 e-family series (out of 3 e-family series - the series of basic bosons (8e8&, 6e6&, ...), the series of maximal-% bosons (8e2468&, 6e246&, ...), and the $e2& series (8e2&, 6e2& ...)) at 2e2& correlates with the role of $\hbar$ in the uncertainty principle.	(23.10)

We discuss a possible correlation between uncertainty and the scale of masses for w-, h-, and o-family bosons

Items including and following item (23.7) provide a possible basis for an uncertainty calculation regarding non-zero-mass basic bosons.

Based on observations, the next item pertains. Here, m(0h1) denotes the mass of a Higgs boson.

$$(m(0h1))^2 / \hbar \approx 0.85 \; (c^5 / G_N) \qquad (23.11)$$

Perhaps, people will find that an expression for mass squared divided by $\hbar$ has appeal, based on item (23.9). (People might say that item (23.9) represents that a square of energy scales with an uncertainty.) Perhaps, people will find that masses for w-, h-, and o-family particles correlate with the values of $\hbar$, c, and G_N.

Perhaps, people will find significance that, in formulas pertaining to a series of lengths [Items following item (6.11)], G_N^{-1} correlates with c^0 and G_N correlates with $\hbar^0$.

We discuss handedness/chirality

Above, we base q- and l-family representations on interactions with o- and w-family basic bosons. [Section 11] For each such representation, there is exactly 1 even integer ї (with 2≤ї≤10), such that, for the Pї-and-P(ї+1) oscillator pair, exactly 1 of $n_{P\ddot{\imath}}$=−1 and $n_{P(\ddot{\imath}+1)}$=−1 pertains. Here, we denote that oscillator (Pї or P(ї+1)) by Pó. Also, we state a concept that, for those q- and l-family representations and for ύ such that 2≤ύ<ї, if $n_{P\acute{\upsilon}}$=−1, the fermion does not interact via e-family modes for which, for the e-family-mode ground state, $n_{P\acute{\upsilon}}$=0 (or, =#).

People might say that such a Pó oscillator is not relevant for the fermion. People might consider the scenario the next items sketch.

Drop, from the representation for the fermion, the Pó oscillator (23.12)
Define #P' = #P − 1 (23.13)
Note that, for the fermion, S = (#P' − 1)/2 (23.14)
- Note that S=(#P'−1)/2 for fermions parallels S=(#P−1)/2 for bosons

Drop, from the representation for the fermion, 1 QE-like oscillator (23.15)
- If the fermion belongs to the q-family, drop the oscillator paired (via an E(χ+1)-and-Eχ oscillator pair for which χ is an even integer) with the oscillator for which −F pertains
  - For quarks, −F = −2
- If the fermion is a lepton,
  - Redo the Section 11 representation to have #E=3, n_{E1}=−3, and either n_{E3}=−1 (and n_{E2}=−2) or n_{E2}=−1 (and n_{E3}=−2)
  - Realize that, for these purposes, n_{E3} and n_{E2} need not pertain to interactions between the fermion and the e-family
  - Drop the QE-like oscillator for which −1 pertains

Note that, for the fermion and this representation, ... (23.16)
- For #E' = #E − 1, people might consider that a form of IOM(#E',#P') pertains, with each of #E' and #P' being even
- Œ=0

People might correlate the evenness or oddness of ó (for the dropped QP-like oscillator) with handedness/chirality for the fermion.

We discuss possibilities regarding black holes and quasars

The next items might reflect some people's thinking regarding black holes and quasars.

While and after a black hole first forms, the density of matter increases based on the influence of 4e4& (23.17)
Eventually, the density becomes large enough that 4e24& dominates 4e4& (23.18)
- Collapse slows
- Collapse may partly reverse
  - For example, non-neutrino fermions may repel each other while neutrinos do not experience much outward acceleration
  - A quasar may develop

Possibly, collapse continues (23.19)
- Eventually, 6e246& may dominate 4e24& and collapse accelerates
- Eventually, 8e2468& may dominate 6e246& and collapse may at least partly reverse
  - A quasar may develop

We discuss possible phenomena people correlate with the term leptoquark

People discuss possibilities for interactions that would convert leptons into quarks or convert quarks to leptons. We use the acronym QTFL to denote quark to or from lepton. People correlate the term leptoquark with (as yet, hypothetical) bosons that might intermediate QTFL reactions. People construct models for characteristics people associate with leptoquarks.

Experimental results provide lower bounds for masses of various hypothesized leptoquarks. The next item notes a range of reported lower bounds (in GeV/c^2 and with minimum confidence levels of 95%). [Ref.23.1]

226 - 685 GeV/c^2 (23.20)

Items including and following item (18.48) show a possible QE-like SU(3)-related symmetry correlating with neutrinos. For the e-family, QE-like SU(3) symmetry correlates with 4e4&. People might say that $o3 and $o2 (for $ = 4, 6, 8, or 10) could add fractional charge to a neutrino. The magnitude of that charge could be |Q'| = 1/3 (for o-type-3) or |Q'| = 1/5, 1/7, 1/9, or 1/11 (for o-type-11). A net addition corresponding to |Q'| = 1/3 would convert the neutrino into a quark. Such an addition could occur (for o-type-3) via any $o3 or $o2 (for $ = 4, 6, 8, or 10) or (for o-type-11) via 3 interactions involving 8o3 or 3 interactions involving 8o2.

Regarding circumstances under which 1 QTFL conversion could occur, people might consider issues related to how many QTFL reactions (at least 3) would need to take place in (in essence) a limited volume of space and a limited period of time. O-family interactions have limited range.

The next items reflect what people might say.

- QTFL reactions convert Ω<0 fermions into Ω>0 fermions (and vice-versa) (23.21)
- IOM correlates with possibilities that o-family bosons can mediate QTFL reactions (23.22)
- O-family related thresholds [items following item (17.29)] do not seem incompatible with lower bounds people discuss for leptoquark masses [item (23.20)] (23.23)
- QTFL reactions might have significant impact in dense material ... (23.24)
  - Early in the big bang
  - In black holes
- Observations have yet to detect or rule out (to some confidence level) QTFL (23.25)

We discuss possibilities regarding classical-physics coordinate systems

The next items might reflect some people's thinking regarding some coordinate systems for classical-physics space time.

To the extent that only maximal-% e-family forces matter, ... (23.26)

- People can represent space time by 1 QE-like coordinate and 3 QP-like coordinates, such that ...
  - The Minkowski metric pertains
  - No curvature pertains
- Such a representation correlates with ... (23.27)
  - all e-family forces for which, in #E=1 representations, n_{E1}=0 for the ground state
  - co-moving coordinates

To the extent that other e-family forces matter, ... (23.28)

- People can add QE-like dimensions in a way such that ...
  - For the 3 QP-like coordinates, flatness pertains
  - Curvature correlates with combinations of QE-like and QP-like coordinates and with combinations of multiple QE-like coordinates

For example, to consider effects of 4e4&, people can use #E=3 (23.29)

- Here, items including and following item (8.33) pertain
- Here, for gravitons, ...
  - n_{E3}=n_{E2}=n_{E1}=0 for the ground state
- People can consider that gravity (as represented in a flat space time with 1 QE-like and 3 QP-like coordinates) includes a QP-like vector-potential-like component as well as a QE-like scalar-potential-like component
  - People might say that, in effect, an analog of Maxwell's equations includes (for the 1 QE-like and 3 QP-like dimensions of a 4-dimension space time) a magnetic-field-like component (even if observers would consider the object generating the gravitational field to be stationary)

Regarding the 4e4& field produced by a star, galaxy, or other object, people might say that ... (23.30)

- The influence of this field on other objects correlates with those other objects' G-masses (PROPE2)
  - A photon has G-mass>0, with a magnitude appropriate to the energy of the photon as observed in the reference frame of an observer of the photon
    - Relative to the object (star, galaxy, etc.), people might consider a frame of reference in which that object is at rest
- The vector-potential-like component correlates with phenomena such as a shift of the perihelion for a planet's orbit
  - PROPE2 applies regarding the planet

The next items might reflect some people's further thinking.

With respect to such a coordinate system ... (23.31)

- Each object that interacts via 4e4& has G-mass≠0 (23.32)
  - For a (K-mass≠0) object moving at a speed v<c, ...
    - $(\text{G-mass})^2 = (\text{K-mass})^2 / (1-(v/c)^2)$
- Perhaps people and computers can easily sum 4e4&-related scalar/vector-potential analogs that correlate with various objects (23.33)
  - Here, the term analogs refers to analogs to the scalar potential and vector potential that are related to 2e2& (and, that people can use to compute 2e2&-related E (or, electric) and B (or, magnetic) fields)
  - Here, people might consider that the analogs may include ...
    - vectors that have more than 4 components
    - tensors that have more than 1 index (possibly, with more than 4 values for each index)
  - Here, the 4e4&-related analogs to the E and B fields might be 3-vectors or might be 4-vectors

We discuss possibilities related to possible IOM for which $D_{*P}\neq 3$

The next item shows a formula pertaining to spin values that could arise for a choice of $D_{*P}=1$.

$\Omega = \pm S(S+D_{*P}-2) = \pm S(S-1)$, with ... (23.34)

- $S(S-1) = 0, 0, 2, 6, \ldots$ respectively for $S = 0, 1, 2, 3, \ldots$
- $S(S-1)$ would be $-1/4$ for $S = 1/2$
- $S(S-1) = 3/4, 15/4 \ldots$ respectively for $S = 3/2, 5/2, \ldots$

The next item shows solutions for $\nu<0$.

$\nu = -1/2$ correlates with edge solutions (23.35)

No $\nu<0$ inside solutions exist (23.36)

We interpret item (23.36) as correlating with non-existence of a field for otherwise potentially possible fermion particles corresponding to item (23.35) and $S\geq 3/2$. Also, we interpret results as correlating with non-existence of $D_{*P}=1$ bosons.

For odd $D_{*P}\geq 5$, Ω cannot match $\pm S(S+1)$ for which 2S is a non-negative integer.

All as-yet-known physics exhibits spins for which $S(S+1)$ pertains with 2S being a non-negative integer.

Apparently, $D_{*P}=3$ correlates with all known observations.

We note some possibly related ratios

The next items pertain.

m(0h1) / m(least neutrino mass or neutrino-mass-related math eigenvalue) (23.37)

$\sim 2\times 10^{20}$

$\beta^6 \sim 2\times 10^{21}$ (23.38)

$Z \sim 1\times 10^{22}$ (23.39)

In this paper, we do not further speculate regarding possible correlations among these 3 numbers.

We note a possible extended symmetry

Per item (13.12), the next item might correlate with a relevant symmetry.

$$SU(17) \supset SU(10)\times U(1)\times SU(7) \quad (23.40)$$

Comments

We suggest research

SOR.23.1 Determine spatial dependence pertaining to s-family-mediated interactions.

SOR.23.2 Determine the extent to which the s-family correlates with the strong interaction's R^0 (asymptotic freedom) spatial dependence.

SOR.23.3 Determine the extent to which o-family bosons provide for the strong interaction's varying from R^0 spatial dependence.

SOR.23.4 Measure spatial dependences for interactions correlating with w-, h-, and o-family bosons.

SOR.23.5 Determine ranges for o-family forces.

SOR.23.6 Detect (or infer) or rule out (to some confidence level) the existence of reactions that convert leptons into quarks or quarks into leptons.

SOR.23.7 Detect (or infer) or rule out (to some confidence level), say for 4e4& and 6e6&, fields analogous to the magnetic field people associate with 2e2&. (Here, we have in mind a space-time coordinate system in which the 3 QP-like dimensions are flat.)

SOR.23.8 Infer or rule out (to some confidence level) that 4e24& and/or 8e2468& produce quasars from matter associated with black holes.

SOR.23.9 To what extent might non-2e2% e-family fields analogous to the magnetic field correlate with jets associated with quasars or with other phenomena? (Here, we have in mind a space-time coordinate system in which the 3 QP-like dimensions are flat.)

STR.23.1 Estimate ranges for o-family forces.

STR.23.2 To what extent might people benefit by considering that the role of ħ in the uncertainty principle correlates with 2e2& belonging to at least 2 of the e-family basic-boson series, the e-family maximal-% series, and the e-family $e2& series?

STR.23.3 Determine the extent to which people might benefit from considering that the masses of w-, h-, and o-family bosons relate to $(\hbar c^5/G_N)^{1/2}$.

STR.23.4 To what extent do masses of q- and l-family fermions correlate with masses of w- and h-family bosons?

STR.23.5 Better relate IOM and handedness/chirality (than does discussion related to item (23.12)).

STR.23.6 Suggest ways to detect (or infer) or rule out (to some confidence level) the existence of reactions that convert leptons into quarks or quarks into leptons.

STR.23.7 To what extent might people benefit by using parallels to the electromagnetic scalar/vector potential when describing gravity and other non-strictly-electromagnetic $e%& interactions? (For example, see items including and following item (23.26) and see items including and following item (23.31).)

STR.23.8 To what extent might people extend the Standard Model to include gravity and non-traditional e-family interactions? (For example, how much can people base such an extension on potentials, currents, and so forth such as items including and following item (23.26) suggest?)

STR.23.9 Discuss implications of SU(17)-related (and higher-dimension SU-related or other) possibly relevant symmetries. (Here, we have in mind discussion related to item (23.40).)

STR.23.10 To what extent might people, based on this work, change estimates of the number of independent physical constants?

STR.23.11 Develop more-coherent or more-compact (than this paper shows) models that produce or improve on results this paper offers.

STR.23.12 Develop theory people can use to produce models (such as this paper shows or STR.23.11 suggests).

STR.23.13 To what extent can people benefit by developing hybrid models that better integrate aspects of IOM and the Standard Model (or other traditional physics)?

STR.23.14 For what other applications might people use IOM?

We list references

Ref.23.1 J. Beringer et al. (Particle Data Group), *PR D86*, 010001 (2012) and 2013 partial update for the 2014 edition (URL: http://pdg.lbl.gov).

Part 6 Perspective

Section 24 Summary and possible further uses

Abs.24.1 We estimate the extent to which this work meets some otherwise perhaps unmet needs.
Abs.24.2 We suggest possible further uses of IOM.

We estimate the extent to which work in this paper meets some otherwise perhaps unmet needs

Items following item (1.1) list needs for which people say traditional mathematical models fall short. The next items estimate extents to which work in this paper meets those needs. For each estimate, we point to least 1 underlying concept.

Needs not met via traditional mathematical physics	Possible advances	
		(24.1)
Provide models people can use to …	Work in this paper may …	(24.2)
• Provide a basis for the Standard Model symmetry SU(3)×SU(2)×U(1)	• Fulfill this need • IOM(3,3) symmetries	(24.3)
• Determine physics-relevant groups that contain SU(3)×SU(2)×U(1)	• Fulfill this need • SU(7)	(24.4)

• List possible basic particles that have not been observed	• Fulfill this need
	  • E-family beyond spin 1
	  • O-family
	  • Q-family beyond spin 1/2
		(24.5)

• Explain the number, 3, of generations of fermions	• Fulfill this need • IOM math	(24.6)
• Describe quantum gravity	• Fulfill this need • E-family phenomena	(24.7)
• Unify quantum gravity and quantum electromagnetism	• Fulfill this need • E-family	(24.8)
• Explain the sizes of some symmetry violations (P, CP, …)	• Point to how to fulfill this need • Interactions that contribute beyond symmetry-violating interactions people correlate with Standard Model physics	(24.9)
• Explain neutrino oscillations	• Point to how to fulfill this need • An interaction that facilitates this phenomenon	(24.10)
• Predict neutrino masses	• Point to a few candidate masses or a few candidate mass-related eigenvalues • Extension of a formula for approximate masses for other S=1/2 fermions	(24.11)

<table>
<tr><th>Needs not met
via traditional mathematical physics</th><th>Possible advances</th><th>(24.1)</th></tr>
<tr><td>Provide models people can use to …</td><td>Work in this paper may …</td><td>(24.2)</td></tr>
<tr><td><ul><li>Interrelate masses of basic particles other than charged leptons</li></ul></td><td><ul><li>Point to approximate relationships between masses of non-zero-mass basic bosons<ul><li>IOM $\nu=-1$ inside solutions</li></ul></li><li>Point to approximate threshold energies for producing o-family groups of multiple particles (some groups may correlate with hypothesized leptoquarks)<ul><li>IOM $\nu=-1$ inside solutions</li></ul></li><li>Point to approximate relationships between masses of S=1/2 basic fermions<ul><li>A formula, based on a periodic-table-like table</li></ul></li></ul></td><td>(24.12)</td></tr>
<tr><td><ul><li>Describe dark matter</li></ul></td><td><ul><li>Point to 2 possibilities for not-mutually-exclusive types of dark matter<ul><li>S=3/2 fermions</li><li>Siblings of baryonic matter (except neutrinos)</li></ul></li></ul></td><td>(24.13)</td></tr>
<tr><td><ul><li>Describe dark energy</li></ul></td><td><ul><li>Point to 2 possibilities for not-mutually-exclusive types of dark-energy stuff<ul><li>S=7/2 fermions</li><li>Peers of the combination of baryonic matter (except neutrinos), neutrinos, and dark matter</li></ul></li><li>Point to 2 possibilities for not-mutually-exclusive types of beyond-dark-energy stuff<ul><li>S=9/2 fermions</li><li>A peer of the combination of baryonic matter (except neutrinos), neutrinos, dark matter, and dark-energy stuff</li></ul></li></ul></td><td>(24.14)</td></tr>
<tr><td><ul><li>Explain changes in the rate of expansion of the universe</li></ul></td><td><ul><li>Point to forces that regulate the rate<ul><li>Maximal-% e-family bosons</li></ul></li></ul></td><td>(24.15)</td></tr>
<tr><td><ul><li>Explain the flatness of the universe ($\Omega_0\approx 1$)</li></ul></td><td><ul><li>Point to forces that seem most influential regarding possible curvature and that seem to correlate with flatness<ul><li>Maximal-% e-family bosons</li></ul></li></ul></td><td>(24.16)</td></tr>
</table>

Needs not met via traditional mathematical physics	Possible advances	(24.1)
Provide models people can use to ...	Work in this paper may ...	(24.2)
• Explain baryon asymmetry (matter/antimatter imbalance)	• Point to interactions that led to this phenomenon • Lasing of o-family bosons	(24.17)
• Address the zero-point energy of the vacuum	• Provide a law that obviates this concern • Œ=0	(24.18)

We note possible other uses for work in this paper

Possibly, people will use IOM to gain new insight regarding the next items.

Traditional topics	Possible uses of IOM	(24.19)
• Nuclear shell model	• People might use IOM to hone or supplant the nuclear shell model	(24.20)
• Harmonic-oscillator math	• People might use IOM for applications (in and beyond physics) this paper does not cover	(24.21)

We think the IOM approach has promise

We hope and think things the next items list. Here, the approach denotes generally work in this paper and specifically IOM.

- People will find value in the approach and results from the approach (24.22)
- People will fix any flaws in the approach and related results (24.23)
- People will find better, more-unified bases for aspects of the approach (24.24)
  - People will develop theory that people can use to produce models
  - Such theory will obviate seeming needs for various guesses
- People will conduct observations and experiments based on people's knowledge of the approach and on results from the approach (24.25)
- People will determine or verify physical numbers, based on such results (24.26)
- People will find better ways to present such theory, the approach, and related results (24.27)

Part 7 Appendices

Section 25 Compendia of section abstracts, guesses, and suggested research

We list statements in section abstracts

Abs.1.1 Traditional mathematical models do not adequately correlate with physics observations.
Abs.1.2 We develop models based on quantum observations and harmonic-oscillator math.
Abs.2.1 IOM (quantum isotropic harmonic oscillator methods) may correlate with and predict physics observations with which traditional models do not correlate.
Abs.3.1 IOM include a concept - QI space - that people might say merges some concepts regarding quantum mechanics, energy-momentum space, and space time.
Abs.4.1 We introduce IOM.
Abs.5.1 We focus on IOM for which $\Omega = \pm S(S+1)$, with S=spin/ħ and with 2S being an integer.
Abs.5.2 IOM correlate with boson and fermion particles and fields.
Abs.6.1 The mass of a tauon may equal a number computed from 4 physics constants.
Abs.6.2 The mass of a tauon may be $1.776814(\sim48)\times10^3$ MeV/c^2.
Abs.6.3 4 physics constants define a series of lengths, including the Planck length.
Abs.7.1 A catalog of families of basic particles points to possible yet-to-be-discovered particles.
Abs.8.1 The e-family includes photons, gravitons, and 2 other zero-mass basic bosons.
Abs.8.2 Each e-family basic boson has 2 modes (or, polarizations).
Abs.8.3 Each of the 4 e-family basic bosons mediates a force with spatial dependence R^{-2}.
Abs.8.4 The e-family includes coherences of the e-family's 4 basic bosons.
Abs.8.5 E-family coherences provide forces with spatial dependences of R^{-4}, R^{-6}, and R^{-8}.
Abs.8.6 IOM may provide a way to avoid dealing with infinite photon ground-state energy.
Abs.9.1 Families of non-zero-mass basic bosons include the w-family (Z, W^-, and W^+ bosons), the h-family (Higgs boson), and the o-family (for which $\Omega<0$).
Abs.10.1 S-family bosons provide for gluons for each of 2 sets of 3 color charges.
Abs.11.1 Families of non-zero-mass basic fermions include the l-family (leptons) and the q-family (for which $\Omega<0$).
Abs.11.2 IOM correlate with leptons, quarks, and related fields.
Abs.11.3 IOM correlate with possible basic fermions with S = 3/2, 7/2, and 9/2.
Abs.11.4 One IOM interpretation correlates with each n-type (or, neutrino-like) basic fermion being its own antiparticle. One IOM interpretation correlates with each basic fermion being distinct from its antiparticle.
Abs.11.5 Each q- or l-family particle is a member of a 3-generation trio.
Abs.12.1 IOM representations correlate with SU(3)×SU(2)×U(1) (or, Standard Model) symmetry for the strong, weak, and electromagnetic interactions.
Abs.12.2 IOM representations can correlate with the statement S=(#P−1)/2 for basic bosons.
Abs.13.1 SU(7) provides a relevant extension of Standard Model symmetry.
Abs.14.1 IOM correlate with kinematics of e- and s-family bosons.
Abs.15.1 IOM correlate with kinematics of basic non-zero-mass particles.
Abs.16.1 E-family coherences provide for changes in the rate of expansion of the observed universe.
Abs.16.2 E-family coherences correlate with the universe's having zero large-scale curvature.
Abs.17.1 IOM correlate with relative masses for w- and h-family bosons.
Abs.17.2 IOM may correlate with masses for o-family bosons.

Thomas.J.Buckholtz@gmail.com http://ThomasJBuckholtz.wordpress.com

Abs.17.3 Threshold energies for creating minimum numbers of o-family bosons may be (in units of mass) 241, 274, 526, ..., 1379 GeV/c^2.
Abs.17.4 The 2o3 boson has charge +(1/3)$|q_e|$ and the 2o2 boson has charge −(1/3)$|q_e|$.
Abs.18.1 A formula approximates masses of quarks and charged leptons.
Abs.18.2 Neutrino masses or neutrino-mass-related math eigenvalues may be, in eV/c^2, approximately 1×10^{-1}, 2×10^{-4}, and 6×10^{-10}.
Abs.19.1 We illustrate interactions involved in fermion-anti-fermion annihilation.
Abs.19.2 We illustrate interactions involved in neutrino oscillations.
Abs.19.3 We illustrate mechanics of channels.
Abs.20.1 Formulas provide approximate relative strengths for interactions mediated by e-family basic bosons and some e-family coherences.
Abs.21.1 We note and interpret observations regarding effects of dark matter and dark-energy stuff.
Abs.21.2 We discuss 2 possible types of DMF (dark-matter basic fermions) - S=3/2 basic fermions and siblings of BMCF (baryonic-matter charged fermions).
Abs.21.3 We discuss 2 possible types of DEF (dark-energy basic fermions) - S=7/2 basic fermions and peers of DM\`N\`BMC\`F (DMF + NF (neutrinos correlating with DMF + BMCF) + BMCF).
Abs.21.4 We discuss 2 possible types of BDEF (beyond-dark-energy basic fermions) - S= 9/2 basic fermions and a peer of NBDEF (DEF + DM\`N\`BMC\`F).
Abs.22.1 Lasing of o-family particles provided key effects leading to matter/antimatter imbalance.
Abs.22.2 E-family coherences provide for phenomena people attribute to axions.
Abs.22.3 O-family particles provide for CPT-related symmetry violations.
Abs.22.4 O-family particles with S≥2 close gaps between magnitudes of violations people estimate via the Standard Model and magnitudes people measure.
Abs.23.1 Ω<0 correlates with an inability for people to observe particles as free-ranging and with particles' exhibiting quantum QP-like behavior.
Abs.23.2 Perhaps, aspects of the uncertainty principle correlate with 2e2& being at least 2 of a basic boson, maximal-% boson, and $e2&-series boson.
Abs.23.3 People may derive from IOM useful insight about handedness/chirality.
Abs.23.4 Maximal-% e-family forces pertain regarding black holes and may produce quasars.
Abs.23.5 IOM may correlate with interactions people say (hypothetical) leptoquarks might mediate.
Abs.24.1 We estimate the extent to which this work meets some otherwise perhaps unmet needs.
Abs.24.2 We suggest possible further uses of IOM.

We list guesses

Gss.4.1 For an edge case with −2ν an even positive integer, 1 type-1 solution exists.
Gss.4.2 For an edge case with −2ν an odd positive integer, 3 orthogonal type-1 solutions exist.
Gss.5.1 Non-traditional IOM having D_{*P}=3 correlate with the basic particles and with some properties of basic particles.
Gss.5.2 For basic particles, ν=−1 correlates with basic bosons and their fields. ν=−3/2 correlates with basic fermion particles. ν=−1/2 correlates with fermion fields.
Gss.5.3 For basic particles, Ω=+S(S+1)>0 correlates with QE-like phenomena, Ω=0 correlates with the Higgs boson, and Ω=−S(S+1)<0 correlates with QP-like phenomena.
Gss.6.1 β' = β.
Gss.6.2 We attach significance to $\lambda_\$$ for which a particle property has an exponent γ=0.
Gss.6.3 Regarding λ_0, people can consider q_e to be a particle property for which $|q_e|^0$ pertains.
Gss.8.1 For e-family members, the force imparted between 2 non-overlapping objects scales as $R^{\acute{\upsilon}}$. Here, $\acute{\upsilon}=2n_{P1}$. Here, R denotes the distance between a center of property of one object and a center of property of the other object.

Gss.8.2 D_E=1, D_P=5 solutions provide a model for gravitons.
Gss.8.3 In the expression $(4/3)(\beta^6)^2 = \{(q_e)^2/(4\pi\varepsilon_0)\} / \{G_N(m_e)^2\}$, the leftmost exponent 2 represents the number of vertices in a Feynman diagram, β^6 represents the ratio of strengths per channel for electromagnetism and gravity (for an interaction between 2 electrons), 4 represents the number of channels for a photon, and 3 represents the number of channels for a graviton.
Gss.8.4 For photons and gravitons, each of the 3 harmonic oscillator pairs P6-and-P7, P8-and-P9, and P10-and-P11 is closed and correlates with a channel.
Gss.8.5 $e%& forces for which % contains a 6 couple to PROPE6.
Gss.8.6 $e%& forces for which % contains an 8 couple to PROPE8.
Gss.9.1 0-family bosons for which (for the ground state) $n_{E1}=0$ transfer PROPO1.
Gss.9.2 0-family bosons for which (for the ground state) $n_{E3}=0$ or $n_{E2}=0$ transfer PROPE2.
Gss.9.3 0-family bosons for which (for the ground state) $n_{E5}=0$ or $n_{E4}=0$ transfer PROPE4.
Gss.9.4 0-family bosons for which (for the ground state) $n_{E7}=0$ or $n_{E6}=0$ transfer PROPE6.
Gss.9.5 0-family bosons for which (for the ground state) $n_{E9}=0$ or $n_{E8}=0$ transfer PROPE8.
Gss.9.6 0-family bosons for which (for the ground state) $n_{EB}=0$ or $n_{EA}=0$ transfer PROPO10.
Gss.9.7 Basic particles for which $\Omega<0$ cannot range freely.
Gss.9.8 0-family basic particles are created in at least triplets.
Gss.10.1 One trio of s-family bosons provides for gluons pertaining to quarks people consider to be matter. The other trio pertains to quarks people consider to be antimatter.
Gss.11.1 For the l-family, combinations of the 4 solutions correspond to 2 of the 3 possible members of an L=1 set (the M=0 member does not apply) and to the 1 member (M=0) of an L=0 set.
Gss.11.2 For the q-family, for S=1/2, combinations of the 4 solutions correspond to 4 of the 5 members of an L=2 set (the M=0 member does not apply).
Gss.11.3 For ï an even positive integer, each of the | $n_{P\ddot{\imath}}$, $n_{P(\ddot{\imath}+1)}$ > states denoted by | −1 , −F > or by | −F , −1 > corresponds to spin/ħ = 1/2. Here, F is an integer and F≥2.
Gss.11.4 For the q-family, for a representation for which there is an even positive integer ï for which the | $n_{P\ddot{\imath}}$, $n_{P(\ddot{\imath}+1)}$ > state denoted by | −1 , −F > or by | −F , −1 > pertains, there is an even positive integer ó for which the | $n_{E(\acute{o}+1)}$, $n_{E\acute{o}}$ > state denoted by | −1 , −F > or by | −F , −1 > pertains.
Gss.13.1 Symmetries related to SU(3), SU(5), and SU(7) pertain for, respectively, 4e4&, 6e6&, and 8e8&.
Gss.15.1 A merger of 2 sets (1 QE-like and 1 QP-like), each of 2 operators, correlates with a standard representation for $E^2-c^2P^2$ that people use based on the Dirac equation. Operators act on 4-component spinors. People can represent aspects of the operators via gamma matrices.
Gss.15.2 W-family masses correlate with Ω_2-related $\nu=-1$ inside solutions.
Gss.16.1 For observed astrophysical objects of above some size, era FEó correlates with era ZEύ, for 1≤ó=ύ≤3.
Gss.16.2 8e2468& repels astrophysical objects from each other. 6e246& attracts astrophysical objects to each other. 4e24& repels astrophysical objects from each other.
Gss.16.3 For observations expressed in terms of co-moving coordinates (with 1 QE-like dimension and 3 QP-like dimensions), people would say that each of the forces 2e2&, 4e24&, 6e246&, and 8e2468& does not contribute to any apparent curvature of space time.
Gss.16.4 To the extent maximal-% e-family bosons dominated (throughout the past history of the universe) interactions between objects, people can consider those objects to be part of a universe for which $\Omega_0 \approx 1$.
Gss.18.1 For charged leptons (either M'=−3 or M'=+3), people can benefit by correlating the range −1≤M"≤3 with an L=2 system.
Gss.18.2 For the L=2 system that includes charged leptons, $m(M'',-3) \propto e^{M''\zeta''}(1+d(M''))$, in which −1≤M"≤3, d(0)=d(2), and d(−1)=d(1)=d(3)=0.
Gss.18.3 The formula for m(M", M') has meaning for M"<−1. The trigonometric-like pattern for d(M") continues throughout the range −6≤M"≤3. d(M",0) = 0.

Gss.20.1 For photon-graviton series basic bosons, for interactions between 2 M"=0 leptons, the relative vertex strength per r/chan follows a pattern established by the relative vertex strengths per r/chan for photons and gravitons.

Gss.20.2 For interactions between 2 electrons, the strengths of 4e4& and 4e24& are equal at a particle separation of roughly λ_4.

Gss.20.3 For interactions between 2 electrons, the strengths of 6e6& and 6e246& are equal at a particle separation of roughly λ_6.

Gss.20.4 For interactions between 2 electrons, the strengths of 8e8& and 8e2468& are equal at a particle separation of roughly λ_8.

Gss.21.1 2 not-mutually-exclusive candidates exist for BDEF. 1 candidate is S=9/2 basic fermions. The other candidate is 4 peers of DM`N`BMC`F.

Gss.21.2 2 not-mutually-exclusive candidates exist for DEF (or, DET2F + DET1F). 1 candidate is S=7/2 basic fermions. The other candidate is 3 peers of DM`N`BMC`F.

Gss.21.3 2 not-mutually-exclusive candidates exist for DMF basic fermions. 1 candidate is S=3/2 basic fermions. The other candidate is 5 ensembles that are siblings of the BMCF ensemble.

Gss.22.1 People consider interactions that convert anti-quarks into quarks (or vice-versa) to violate P symmetry.

Gss.22.2 Phenomena people associate with axions exist. People can associate such phenomena with e-family coherences.

Gss.22.3 Differences between ύ-symmetry for IOM(11,11) and ύ-symmetry for IOM(3,3) correlate with sizes of ύ-violations people do not associate with Standard Model physics. Here, ύ can be (at least) C, P, or T.

Gss.23.1 In the sense of traditional quantum perturbation models for interactions between elementary particles, people might consider that q- and o-family particles traverse space-like trajectories between vertices.

Gss.23.2 Item (23.1) correlates with an inability to observe free-ranging instances of $\Omega<0$ basic particles.

Gss.23.3 The confluence of at least 2 e-family series (out of 3 e-family series - the series of basic bosons (8e8&, 6e6&, ...), the series of maximal-% bosons (8e2468&, 6e246&, ...), and the $e2& series (8e2&, 6e2& ...)) at 2e2& correlates with the role of ħ in the uncertainty principle.

We list suggestions for observational and experimental research

SOR.6.1 Verify (to a smaller than current experimental uncertainty-range) or refute $\beta' = \beta$ and the predicted tauon mass.

SOR.8.1 Detect instances or effects of, or rule out (to some confidence level) the existence of, 4e24& coherences.

SOR.8.2 Determine the extent to which 4e24& includes mixed (even and odd) polarization modes.

SOR.8.3 Conduct experiments to produce or rule out (to some confidence level) reactions that would produce 4e24& from multiply excited 2e2& modes.

SOR.8.4 Measure or infer S for various e-family coherences and various other non-basic-boson e-family members.

SOR.8.5 Determine (to some confidence level) or rule out that each basic e-family boson in a coherence moves in a direction equal to the direction for each other basic e-family boson in the coherence.

SOR.8.6 Measure or infer signs and magnitudes for forces mediated by e-family members other than 2e2& and 4e4&.

SOR.9.1 Verify or rule out (to some confidence level) existence of o-family bosons.

SOR.9.2 Verify or rule out (to some confidence level) that o-family bosons cannot be created singly.

SOR.9.3	Verify or rule out (to some confidence level) changes to nuclear theory people propose based on o-family physics.
SOR.11.1	Detect (or infer) or rule out (to some confidence level) the existence of basic fermions for which S = 3/2, 7/2, or 9/2.
SOR.11.2	Rule out (to some confidence level) or detect the existence of basic fermions for which S = 5/2 or 11/2.
SOR.11.3	Measure or infer properties of S≥3/2 basic fermions.
SOR.11.4	Measure or infer reactions in which S≥3/2 basic fermions participate.
SOR.11.5	Verify or rule out (to some confidence level) q- and l-family interaction rules we show regarding the w-, h-, and o-families. Determine strengths for interactions for which strengths are yet to be determined.
SOR.11.6	To what extent does either n-type model (n-type-S or n-type-D) better correlate with observations than does the other n-type model?
SOR.13.1	Verify (to some confidence level) or refute that SU(7) symmetry correlates with observations.
SOR.16.1	Estimate charges of objects for which 4e24& currently dominates.
SOR.17.1	Determine properties (such as charge, mass, and magnetic moment) of o-family bosons.
SOR.17.2	Detect (or infer) or rule out (to some confidence level) the existence of o-family bosons with charges Q' (in units of $\lvert q_e \rvert$) of ±1/3 and 0,
SOR.17.3	Detect (or infer) or rule out (to some confidence level) the existence of o-family bosons with Q' of ±1/5, ±1/7, ±1/9 and ±1/11.
SOR.17.4	Measure masses for basic-boson o-family members and/or measure threshold energies for compound particles based on o-family members.
SOR.17.5	Determine or rule out (to some confidence level) non-zero binding energies for at-least-triplets of o-family bosons.
SOR.17.6	Verify or rule out (to some confidence level) that much of the difference between the W-boson mass we calculate and the observed W-boson mass correlates with a non-zero magnetic moment for W bosons.
SOR.18.1	Measure neutrino masses.
SOR.19.1	Determine dependences of neutrino-oscillation rates on influences of mass or gravity.
SOR.21.1	Find or rule out (to some confidence level) evidence (other than currently assumed evidence) of effects on baryonic matter of dark matter.
SOR.21.2	Detect (or infer) or rule out (to some confidence level) that 4e24% mediates interactions between BMCF and DMF. (Here, we have in mind that 4e4& mediates interactions between BMCF and DMF and that 2e2& does not mediate direct interactions between BMCF and DMF.)
SOR.21.3	Measure or infer masses for (or rule out (to some confidence level) existence of) S≥3/2 basic fermions.
SOR.21.4	Determine the extents to which fields related to various 4e%&, 6e%&, and 8e%& bosons have impacted CMB radiation.
SOR.21.5	Determine or rule out (to some confidence level) the existence of more than 1 ensemble.
SOR.21.6	Determine the extents to which fermion properties vary by ensemble.
SOR.21.7	Infer w-, h-, and o-family interaction strengths and/or big-bang phenomena related to w-, h-, and o-family-mediated interactions.
SOR.21.8	Detect (or infer) or rule out (to some confidence level) effects on NBDEF of BDEF.
SOR.21.9	Determine the extent to which observations (for example, about the uniformity of distribution of NBDEF) cannot be explained by traditional physics and/or work in this paper directly related to NBDEF. Estimate the extent to which people might attribute any such unexplained phenomena to the presence of BDEF (for example, via effects early in the big bang).

SOR.21.10 Determine the extent to which (DMF, DET1F, DET2F, and/or BDEF) ensembles contribute to actual densities of the universe. (Here we differentiate actual from inferred {for example, inferred via CMB data}.)

SOR.22.1 Detect or rule out (to some confidence level) that 2o3 and 2o2 bosons can covert a q-family S=1/2 fermion from anti-quark to quark and vice-versa.

SOR.22.2 Determine the extents to which w- and o-family-mediated interactions and/or e-family-mediated interactions account for observed P violations, CP violations, or other such violations.

SOR.23.1 Determine spatial dependence pertaining to s-family-mediated interactions.

SOR.23.2 Determine the extent to which the s-family correlates with the strong interaction's R^0 (asymptotic freedom) spatial dependence.

SOR.23.3 Determine the extent to which o-family bosons provide for the strong interaction's varying from R^0 spatial dependence.

SOR.23.4 Measure spatial dependences for interactions correlating with w-, h-, and o-family bosons.

SOR.23.5 Determine ranges for o-family forces.

SOR.23.6 Detect (or infer) or rule out (to some confidence level) the existence of reactions that convert leptons into quarks or quarks into leptons.

SOR.23.7 Detect (or infer) or rule out (to some confidence level), say for 4e4& and 6e6&, fields analogous to the magnetic field people associate with 2e2&. (Here, we have in mind a space-time coordinate system in which the 3 QP-like dimensions are flat.)

SOR.23.8 Infer or rule out (to some confidence level) that 4e24& and/or 8e2468& produce quasars from matter associated with black holes.

SOR.23.9 To what extent might non-2e2% e-family fields analogous to the magnetic field correlate with jets associated with quasars or with other phenomena? (Here, we have in mind a space-time coordinate system in which the 3 QP-like dimensions are flat.)

We list suggestions for theoretical research

STR.4.1 Complete mathematics related to type-1 solutions sufficiently to describe wave functions for edge cases for $D_P = 3$ and $\nu = -3/2$. (Possibly, extend the work to pertain to edge cases for other D_P and ν.)

STR.4.2 To what extent might people derive benefit from IOM for any 1 or more than 1 of the following? D can be an integer < 1. D can be other than an integer. 2ν can be other than an integer. $\pm_\chi$ can be other than +1 or −1.

STR.4.3 To what extent would it be useful, for D=3, for people to consider a quantum number s' such that S=(s'−1)/2? (Here, $S(S+1)=(1/4)\cdot((s')^2-1)$. Here, perhaps, s' is any non-zero integer.)

STR.4.4 Explore IOM for cases in which D_E is even and D_P is even.

STR.4.5 How might people improve or extend the technique for cataloging quantum approaches?

STR.5.1 To what extent would people benefit from defining and using a value for a D_{*E}?

STR.6.1 To what extent does a relationship for m_{muon}/m_e involving physical constants exist? (Here we have in mind a parallel to item (6.46) and not item (18.100), the Koide formula.)

STR.8.1 Predict the extent to which e-family coherences exhibit mixed (even and odd) polarization modes.

STR.8.2 Harmonize models and observations or experiments regarding S for e-family members.

STR.8.3 To what extent might people find it appropriate to associate $(t')^0$ behavior with $e%&-mediated interactions? (Here, t' denotes time. [Section 14])

STR.9.1 How best might people explore the existence and characteristics of $o% particles?

STR.9.2 What known or new phenomena people might explain based on the o-family?

STR.9.3 To what extent do o-family bosons correspond to aspects of the shell model for atomic nuclei? (Harmonic-oscillator math seems to pertain to each of the o-family and the shell model.)

STR.9.4 To what extent might people explain properties of atomic nuclei, based on o-family bosons (and gluons and other physics)?

STR.9.5 To what extent might people explain properties of neutron stars, based on o-family bosons (and other physics)?

STR.10.1 To what extent might people benefit by considering the possibility that the s-family mediates a force with R^0 spatial dependence? (Here, we have in mind asymptotic freedom.)

STR.11.1 Estimate properties PROPE2, PROPE4, PROPE6, PROPE8, and PROPO10 for basic fermions for which $S \geq 3/2$.

STR.11.2 Estimate interaction strengths for interactions between w-, h-, and o-family bosons and $S \geq 3/2$ basic fermions.

STR.11.3 Describe possible composite objects (such as nuclei or atoms) for which $S \geq 3/2$ basic fermions would be components.

STR.11.4 To what extent might people benefit by considering that generation-2 and generation-3 fermions might not be truly basic particles (and are excitations of generation-1 basic particles)?

STR.13.1 Better connect theoretically 2 concepts (channels and the non-applicability of a possible SU(9)-related symmetry) that point to a limit of $S \leq 4$ for the e-family.

STR.13.2 To what extent might people benefit by considering SU(5) to be a relevant symmetry larger than the Standard Model symmetry?

STR.15.1 Show that IOM correlate with applicability, regarding basic fermions, of the Dirac equation.

STR.15.2 Determine the extent to which people can benefit by considering $E^2 - c^2P^2$ to be negative for the s- and q-families.

STR.15.3 Determine the extent to which people can benefit by considering $E^2 - c^2P^2$ to both positive and negative for the h-family.

STR.17.1 Estimate a value for the magnetic moment of W bosons.

STR.17.2 To what extent might people benefit by noticing that 2o% masses may equal counterpart 2w% masses and/or that 8o1, 10o1, and 0h1 masses may equal each other?

STR.18.1 To what extent does the appearance, in a formula for the ratios of masses of charged leptons, of a power of the ratio of the square roots of 2 lepton masses correlate with the possible applicability of the Koide formula? [Items (18.99) and (18.100)]

STR.18.2 Determine the extents to which numbers we state as correlating with neutrinos represent mass-related math eigenvalues and/or neutrino masses. [Items following item (18.89)]

STR.19.1 Determine the extent to which people might benefit by further exploring numbers and mechanics related to channels.

STR.20.1 Develop theory sufficient to predict choices - attraction, repulsion, or neither - for each e family interaction between 2 particles.

STR.20.2 Predict strengths and directions (attraction or repulsion) for e-family forces other than 2e2& and 4e4&.

STR.20.3 To what extent do neutrinos interact with \$e%&-for-which-2∈% bosons based on, for example, neutrinos being transformed into virtual pairs, each consisting of a charged lepton and a 2w2 or 2w3?

STR.20.4 Estimate the Hubble constant.

STR.20.5 Develop a suitable IOM perturbation theory (possibly based on something like Feynman diagrams) for e-family interactions.

STR.20.6 Use such an IOM perturbation theory [STR.20.5] to estimate magnetic-moment anomalies. [Items (6.44) and (6.45)]

STR.21.1 How best should people try to directly detect matter not associated with baryonic matter?

STR.21.2 Develop models people can use to estimate masses for S≥3/2 basic fermions.
STR.21.3 Estimate ratios of densities of the universe for baryonic matter, dark matter, dark-energy stuff, and beyond-dark-energy stuff, assuming that S≥3/2 basic fermions provide all DMF fermions, all DET2F fermions, and all BDEF fermions.
STR.21.4 To what extent might people benefit by considering the possibility that an 8-fold symmetry among 8e8&, 6e6&, and 4e4& correlates with the possibility that nature does not exhibit negative values of each of PROPE8, PROPE6, and PROPE4? [Item (21.17)]
STR.21.5 Review observational data and calculations of the density of baryonic matter (equivalent to a mass density of approximately 9.9×10^{-30} g/cm^3, which is similar to mass density of 5.9 protons per cubic meter) to find a basis for determining the extent to which S≥3/2 fermions might pertain. [Ref.21.3]
STR.21.6 For each of the various eras (ZEύ, per Section 16), which interactions most affect neutrinos?
STR.22.1 To what extent do e-family-coherence-mediated interactions correlate in strength with measured CP violations?
STR.22.2 To what extent do $o%-mediated interactions correlate in strength with measured C, P, and T violations?
STR.22.3 To what extent do some o-family-mediated interactions correlate in strength with C, P, and T violations for which the Standard Model can account?
STR.23.1 Estimate ranges for o-family forces.
STR.23.2 To what extent might people benefit by considering that the role of ħ in the uncertainty principle correlates with 2e2& belonging to at least 2 of the e-family basic-boson series, the e-family maximal-% series, and the e-family $e2& series?
STR.23.3 Determine the extent to which people might benefit from considering that the masses of w-, h-, and o-family bosons relate to $(\hbar c^5/G_N)^{1/2}$.
STR.23.4 To what extent do masses of q- and l-family fermions correlate with masses of w- and h-family bosons?
STR.23.5 Better relate IOM and handedness/chirality (than does discussion related to item (23.12)).
STR.23.6 Suggest ways to detect (or infer) or rule out (to some confidence level) the existence of reactions that convert leptons into quarks or quarks into leptons.
STR.23.7 To what extent might people benefit by using parallels to the electromagnetic scalar/vector potential when describing gravity and other non-strictly-electromagnetic $e%& interactions? (For example, see items including and following item (23.26) and see items including and following item (23.31).)
STR.23.8 To what extent might people extend the Standard Model to include gravity and non-traditional e-family interactions? (For example, how much can people base such an extension on potentials, currents, and so forth such as items including and following item (23.26) suggest?)
STR.23.9 Discuss implications of SU(17)-related (and higher-dimension SU-related or other) possibly relevant symmetries. (Here, we have in mind discussion related to item (23.40).)
STR.23.10 To what extent might people, based on this work, change estimates of the number of independent physical constants?
STR.23.11 Develop more-coherent or more-compact (than this paper shows) models that produce or improve on results this paper offers.
STR.23.12 Develop theory people can use to produce models (such as this paper shows or STR.23.11 suggests).
STR.23.13 To what extent can people benefit by developing hybrid models that better integrate aspects of IOM and the Standard Model (or other traditional physics)?
STR.23.14 For what other applications might people use IOM?

Section 26 References

Ref.3.1 John R. Gribbin and Mary Gribbin, *Richard Feynman, A Life In Science*, Dutton, 1997, page 178.

Ref.4.1 Wolfram Alpha, computational knowledge engine, Wolfram Alpha LLC, http://mathworld.wolfram.com/DeltaFunction.html.

Ref.6.1 T. Quinn et al, Improved Determination of G Using Two Methods, *Phys. Rev. Lett,* 111, 101102, 2013. (http://link.aps.org/doi/10.1103/PhysRevLett.111.101102)

Ref.6.2 J. Beringer et al. (Particle Data Group), *Phys. Rev. D86*, 010001 (2012). (http://pdg.lbl.gov/2012/reviews/rpp2012-rev-phys-constants.pdf)

Ref.6.3 J. Beringer et al. (Particle Data Group), *Phys. Rev. D86*, 010001 (2012). (http://pdg.lbl.gov/2012/tables/rpp2012-sum-leptons.pdf)

Ref.6.4 Particle Data Group, Electroweak (web page), *The Particle Adventure*, Lawrence Berkeley National Laboratory, http://www.particleadventure.org/electroweak.html.

Ref.6.5 G. T. Adylov, et. al., A measurement of the electromagnetic size of the pion from direct elastic pion scattering data at 50 GeV/c, *Nuclear Physics B*, Volume 128, Issue 3, 3 October 1977, pages 461-505. (http://dx.doi.org/10.1016/0550-3213(77)90056-6)

Ref.13.1 John Baez and John Huerta, The Algebra of Grand Unified Theories, *Bulletin of the American Mathematical Society, Volume 47*, Number 3, July 2010, pages 483-552. http://www.ams.org/journals/bull/2010-47-03/S0273-0979-10-01294-2 /

Ref.14.1 Wolfram Alpha, computational knowledge engine, Wolfram Alpha LLC, http://mathworld.wolfram.com/Laplacian.html.

Ref.16.1 N. G. Busca, et. al., Baryon Oscillations in the Lyα forest of BOSS quasars, arXiv:1211.2616 [astro-ph.CO].

Ref.16.2 A. Riess, et. al., Type Ia Supernova Discoveries at z > 1 from the *Hubble Space Telescope*: Evidence for Past Deceleration and Constraints on Dark Energy Evolution, *The Astrophysical Journal*, 607, 665 (2004). (doi:10.1086/383612) (http://iopscience.iop.org/0004-637X/607/2/665)

Ref.16.3 N. Gnedin, Cosmological Calculator for the Flat Universe. (http://home.fnal.gov/~gnedin/cc/)

Ref.16.4 NASA, http://map.gsfc.nasa.gov/universe/uni_shape.html

Ref.17.1 J. Beringer et al. (Particle Data Group), *PR D86*, 010001 (2012) and 2013 partial update for the 2014 edition (URL: http://pdg.lbl.gov). (http://pdg.lbl.gov/2013/tables/rpp2013-sum-gauge-higgs-bosons.pdf)

Ref.17.2 CMS collaboration (2012). "Observation of a new boson at a mass of 125 GeV with the CMS experiment at the LHC". *Physics Letters B* 716 (1): 30–61. arXiv:1207.7235. Bibcode:2012PhLB..716...30C. doi:10.1016/j.physletb.2012.08.021.

Ref.17.3 ATLAS collaboration (2012). "Observation of a New Particle in the Search for the Standard Model Higgs Boson with the ATLAS Detector at the LHC". *Physics Letters B* 716 (1): 1–29. arXiv:1207.7214. Bibcode:2012PhLB..716....1A. doi:10.1016/j.physletb.2012.08.020.

Ref.18.1 J. Beringer et al. (Particle Data Group), *Phys. Rev. D86*, 010001 (2012). (http://pdg.lbl.gov/2012/tables/rpp2012-sum-quarks.pdf)

Ref.18.2 J. Beringer et al. (Particle Data Group), *PR D86*, 010001 (2012) and 2013 partial update for the 2014 edition (URL: http://pdg.lbl.gov). (http://pdg.lbl.gov/2013/tables/rpp2013-sum-leptons.pdf)

Ref.18.3 S. Thomas, F. Abdalla, and O. Lahav, Upper Bound of 0.28 eV on the Neutrino Masses from the Largest Photometric Redshift Survey, *Phys. Rev. Lett. 105*, 031301, 2010. (http://arxiv.org/abs/0911.5291)

Ref.18.4 A. Melchiorri, Constraints on Neutrino Physics from Planck, European Space Agency, http://www.rssd.esa.int/SA/PLANCK/docs/eslab47/Session06_CMB_Cosmology_and_Fundamental_Physics/47ESLAB_April_04_17_30_Melchiorri.pdf.

Ref.19.1 J. Beringer et al. (Particle Data Group), *Phys. Rev. D86*, 010001 (2012). "13. Neutrino mass, mixing, and oscillations," page 46. (http://pdg.lbl.gov/2012/reviews/rpp2012-rev-neutrino-mixing.pdf)

Ref.21.1 J. Beringer et al. (Particle Data Group), *Phys. Rev. D86*, 010001 (2012). (http://pdg.lbl.gov/2012/reviews/rpp2012-rev-cosmic-microwave-background.pdf)

Ref.21.2 Mark Peplow, Planck telescope peers into primordial universe, *Nature News*, Nature Publishing Group, March 21, 2013. (http://www.nature.com/news/planck-telescope-peers-into-primordial-universe-1.12658)

Ref.21.3 Wilkinson Microwave Anisotropy Probe, http://wmap.gsfc.nasa.gov/universe/WMAP_Universe.pdf

Ref.23.1 J. Beringer et al. (Particle Data Group), *PR D86*, 010001 (2012) and 2013 partial update for the 2014 edition (URL: http://pdg.lbl.gov).

www.ingramcontent.com/pod-product-compliance
Lightning Source LLC
Chambersburg PA
CBHW080257180526
45167CB00006B/2559